建筑立场系列丛书 No.76
Public Space ReConfigure
公共建筑改造
[意]福克萨斯建筑设计事务所等 | 编
王京 于风军 孙探春 杜丹 罗茜 王晴 | 译
大连理工大学出版社

C3 建筑立场系列丛书 No.76

公共建筑改造

Public Space ReConfigure

公共建筑改造

Space
figure

全球变暖时代的公共空间

尽管美国最近有一位非常有影响力的反气候变化论者当权，但所有的可靠信息都显示地球已经进入了一个新的地质时期，保罗·克鲁岑称之为人类世。人造气体排放到大气中两个世纪之后，气温逐渐升高，极端暴风雨天气加剧，主要城市地区受到雨水过多或者过少的威胁。飓风卡特里娜摧毁了新奥尔良，而达卡每年洪水泛滥则使公共空间变得更加稀少。太平洋基里巴斯岛岛上 10 万居民则成为首批气候变暖的受害者，海平面上升淹没了他们的土地。在设计和规划新的城市或者重建城市的过程中，我们很少考虑全球变暖对公共空间造成的影响。然而不幸的是，在 21 世纪，随着海平面上升 0.2m 至 2m 甚至更多，世界上多达 70% 的大城市将遗憾地发现其公共空间受到威胁。

说到当代的公共空间，主要会涉及犯罪、人与人之间的冷漠和机动车交通问题，简·雅各布斯在《美国大城市的生与死》(1961 年) 中、威廉·H. 怀特在《小城市空间的社会生活》(1980 年) 中以及扬·盖尔在《公共空间·公共生活》(1996 年) 中都谈论过这些问题。他们提出了暂时解决上述问题的方法，如把街道变窄使小汽车无法通过，规划多条带有公共景点的人行通道，设计一些有长椅、喷泉和植被的舒适区以及安装电子警察监视点等。这些暂时解决问题的方法在成功规划现代空间中发挥了价值和作用，如博堡蓬皮杜中心 (皮亚诺和罗杰斯，1976 年)、巴塞罗那兰布拉大街 (Carles Diaz、Xavier Sust、Oscar Tusquets 和 Lluís Clotet，1999 年)

Public Space in the Age of Global Warming

Although a very influential climate-change-denier recently assumed power in the USA, all knowledgeable sources agree that the planet has entered a new geological phase, defined by Paul Crutzen as the Anthropocene. After two centuries of intense human-produced gases released into the atmosphere, the climate is incrementally warming up, freak storms intensifying, and either excess water or lack of it threatens major urban areas. Hurricane Katrina devastated the New Orleans, while the annual inundations in Dhaka make the option of public space tenuous. The 100,000 inhabitants of the Kiribati Islands in the Pacific will be the first climate victims to surrender their land to rising waters. In the design and programming of new or restructured urban areas the effect of global warming on public spaces is rarely considered. Yet as many as 70% of the world's great cities will unfortunately find their open spaces threatened as the waters rise from 0.2 to 2 meters or more during the 21st century.

When considering contemporary public space, the issues addressed by Jane Jacobs (*The Death and Life of Great American Cities, 1961*), William H. Whyte (*The Social Life of Small Urban Spaces, 1980*) and Jan Gehl (*Public Spaces, Public Life, 1996*), mostly involved crime, apathy, and motor traffic. Palliative solutions, such as narrowing streets to discourage cars, planning multiple pedestrian access points with a series of desirable goals in public view, designing comfortable zones with benches, fountains and vegetation, and installing soft-cop surveillance points, proved their merit in planning successful modern spaces such as Place Beaubourg at Center Pompidou (Piano & Rogers, 1976), the Rambla del Raval in Barcelona (Carles Diaz, Xavier Sust, Oscar Tusquets and Lluís Clotet, 1999), or the Piazza Gae Aulenti in Milan (Pelli Clark Pelli, 2013). Adherence to the theory of social triangulation – a

以及米兰 Gae Aulenti 广场 (Pelli Clark Pelli 建筑师事务所, 2013 年) 。遵守社会三角关系理论——一个项目至少具有三种不同的功能, 也为公共空间相互交叉找个借口——对不久的将来来说可能是个好主意, 虽然这样做人们会受到影响, 甚至可能与专注于看手机的人发生冲突。

21 世纪, 手机在生活中的盛行成为影响公共空间未来的另一个不可阻挡的因素, 其对未来公共空间产生的影响可能比气候变化带来的影响更大。虚拟世界的变化速度好像比涨潮的速度还快, 覆盖了人们聚集的公共空间。目前的数据显示, 每 100 个人当中, 96 个人都有手机; 而在 15 年前, 每 100 个人当中, 有手机的仅有 17 人。这告诉所有人一个明确的事实: 手机控制了我们的生活。在世界各地, 公共机构和私营企业都以在公共空间拉网线提供免费 Wi-Fi 服务为荣, 这项服务变得比街头卫生、城市照明、治安维护等还重要。更重要的是, 这种广泛的联系正在对人类产生更大的影响。数字入侵并不是将人们排除在公共空间之外, 而是使他们无论是否在公共场所中都不受空间的限制。作家雪莉·图克尔贴切地把这种现象称为"一起孤独" (2012 年) 。当今在城市中到处都是 Wi-Fi, 并且通过 Wi-Fi, 人与人之间的社交关系被虚拟社交媒体所取代。或许当代公共空间的理想选择是提供"禁止 Wi-Fi"专区而不是提供"免费 Wi-Fi"区域, 就像指定的吸烟区, 它们已经成为最活跃的城市区域。

program with at least three different functions creating pretexts to cross public space – will probably remain a good idea for the near future, even if one has to get his feet wet in doing so, or even more likely collide with someone concentrating on his cell phone messages.

The prevalence of cell phones in 21st-century life introduces another inexorable factor pecking at the future of public space that may have even greater repercussions than climate change. Virtualism seems to be racing faster than the rising tides to cover the spaces where people congregate. The current statistic of 96 cell phones contracting per 100 people, versus that of 17 to 100 only 15 years ago tells the story evident to all: phones are controlling us. Throughout the world public agencies and private businesses have proudly been cabling to guarantee free Wi-Fi in public space, a service that has become more important than street cleaning, lighting, police protection, and so on. More importantly, universal connection is exerting evolutionary consequences on humanity. The digital invasion does not necessarily exclude people from public space, but rather subjects them to despatialization, whether they are in public or not. One author, Julie Turkle, aptly calls this phenomenon "Alone Together" (2012). It is next to impossible to be in urban settings today without witnessing mass Wi-Fi addiction, and with it the displacement of person to person social relations by virtual social media. Perhaps a desirable option for contemporary public space, would be to create sanctuary areas that are "Wi-Fi free" instead of the "free Wi-Fi", something like the designated smoking areas, which have become the most lively urban spaces.

During the 1980s it was quite evident that major political and financial interests, despite the lessons of Jane Jacobs and others, intentionally attempted to reduce the use of public space, for questions of maintenance and

在 20 世纪 80 年代，尽管简·雅各布斯等人著书立说总结了许多经验教训，但出于重大的政治和经济利益考虑，打着关注维护与安全问题的旗号故意减少公共空间的企图仍十分明显。一个明显的例子就是位于辛辛那提市中心的宝洁总部的两个景观广场（科恩·彼得森·福克斯，1982—1985 年）。尽管其校园般的外观十分具有吸引力，带有美丽的绿廊和葡萄藤，但它仍然是一个专有的私人空间，被兢兢业业的保安守卫着，禁止不相关的人员进入。设计也能激发反社会的效应，阿尔多·罗西设计的位于佩鲁贾的 Bacio 广场就是如此。该项目开始于 1983 年，竣工于 20 世纪 90 年代，其原址是制作 Bacio Perugina 巧克力的工厂。虽然它在形状、规模和特点上都可算得上是一个好的意大利广场，并且靠近中央火车站，看起来应该很有吸引力，但是它一直闲置着，有很多深影，就像建筑师所欣赏的乔治·德·基里科的画一样。人们不愿进入这个广场的一个原因是它被抬高了 20 个台阶，限制了人们的视野；另外一个原因是周边建筑物的主入口没有一个是对着广场的。与此同时，在西边政府大楼底层，规划的商店都深陷在建筑物的底部，几乎看不到，所以这些商店也很快就停业了。这些情况属于“城市无人”综合征。当今，出现广场无人的情况部分原因是日益严峻的环境变化造成的，但更主要的却是因为盛行的数字化渗透已经侵入了真实空间。

尽管有这样的阻碍，但是很多人仍然想去公共空间见面会友、购物、游玩、使用手机，必要的时候去示威游行。在过去的 20 年中，文化机构，特别是博物馆、剧院和图书馆，在营造公共空间方面成为主角。这让人想起弗兰克·盖里设计的位于毕尔巴鄂的古根海姆博物馆（1997 年），或伦佐·皮亚诺设计的位于罗马的 Parco della Musica 音乐馆（2002 年，由三个音乐厅组成）的

security. A clear case is the two-block landscaped plaza provided by Procter & Gamble Headquarters in downtown Cincinnati (Kohn Pedersen Fox, 1982-85). Despite its attractive campus-like appearance, with lovely pergolas and vines, it remains an exclusive private space, assiduously guarded by private police to keep all unwanted users out. Such an anti-social effect can be provoked by design as well, seen in Aldo Rossi's Piazza del Bacio in Perugia, a space planned in 1983 and realized in the 1990s on the site of the factory where Bacio Perugina chocolates were once made. Although it has the shape, scale, and characteristics of a good Italian piazza, and its proximity to the central train station would seem a certain attraction, it is consistently empty – full of deep shadows like the Giorgio de Chirico paintings admired by the architect. One thing that keeps people away is that it is raised 20 steps above grade, which limits visual access; another is that none of the principal entries to the buildings give on to the piazza. Meanwhile the shops planned for the ground level of the government building on the western edge were tucked deep into the base of the building, nearly impossible to see, and they soon went out of business. Such episodes belong to the "cities-without-people" syndrome. Today this seems to be occurring by default partly due to impending environmental changes but mostly because of the prevalence of the digital infiltration of real space.

Despite such deterrents many people still desire to go into public space to see each other, to shop, to play, to use their cell phones, and when necessary to demonstrate. During the last 20 years cultural institutions, especially museums, theaters, and libraries, have been the major protagonists in generating public spaces. One thinks of the entry to Frank Gehry's Guggenheim Museum in Bilbao (1997), or the amphitheater-shaped piazza

圆形广场，再或者是位于巴黎的法国国家图书馆（多米尼克·佩罗，1994 年）那卓越超群的露天平台。但是这一代的公共空间功能上仍然很单一，与人们日常生活关系不大，看起来更像是临时展示区或接待区而非人们的聚集之所。

气候变化的到来让人们不得不重新考虑不久的将来公共空间的功能。有明确的迹象表明，全球在着手应对气候变化：在 2015 年 12 月巴黎召开的第 21 届联合国气候变化大会（COP21）上提议并于 2016 年 4 月 22 日"地球日"由 175 个国家签署的协定指出，各方有责任把气温升高的幅度控制在 2°C 之内，最好是控制在 1.5°C 之内。大多数人认为这并不可行。在过去的两个世纪里，二氧化碳碳酸和甲烷气体已经对地球上的冰川和极地冰盖造成了不可逆转的破坏，冰川和冰盖已经减少了三分之一以上，造成了水位升高，暴雨增加，导致海岸线发生明显的变化。2016 年 6 月巴黎发生的洪涝灾害似乎是对 COP21 人们所关注的问题的回应，也说明了公共空间有多么脆弱。

在未来部分被淹没的城市里还可能有公共空间的存在吗？通常的预测是，到 2050 年，地球上 70% 的人口会居住在城市，但是这一预测并没有考虑上述水力失衡问题。在不到一个世纪的时间里，威尼斯的水位上涨了 28cm，经常遭受洪涝灾害，人们开始了两栖生活模式。在涨潮期间，威尼斯主要的公共空间都铺上了 1m 高的木板。从长远的计划来看，解决洪涝的传统办法是抬高堤坝。然而最近，这种"聪明"的方法受到了"摩西工程"的挑战，该方法是在泻湖的三个湖口处筑三座巨大的可调节的堤坝。

of Renzo Piano's Parco della Musica for three concert halls in Rome (2002), or the sublime deck of the National Library of France in Paris (Dominique Perrault, 1994). But the public spaces of this generation, remain monofunctional and marginal to daily life functions and seem more like spaces of momentary display or reception rather than places to gather.

The advent of climate change may offer a compelling reason to rethink the function of public space in the near future. There are clear signs that the world is getting ready: the UN's COP21 statement compiled in Paris in December 2015, and signed by 175 nations on Earth Day April 22, 2016, broadcasts the commitment to arrest climate increase to below 2 degrees, specifically to 1.5 degrees centigrade, which by most accounts does not seem feasible. The impact of CO_2, carbonic acid and CH_4 methane gas during the past 2 centuries has irreversibly damaged the planet's glaciers and polar icecaps which have receded by over one third. This has influenced the rise in water levels and the increase of torrential rainfall, leading to already perceptible changes in shorelines. The floods in Paris in June, 2016, seem to have echoed the concerns of COP21 and demonstrate how fragile public space can be.

Will public space in the partially submerged urban environment of the future be possible? The often cited prediction that 70% of the planet's population will be urban by 2050 does not seem to take into account this hydraulic imbalance. Venice, where the water has risen 28 centimeters in less than a century and which experiences regular inundations, resorts to a model of amphibious adaptation. During periods of acqua alta networks of one-meter-high planks are stretched out over her major public spaces. In terms of long-term plan, the tradi-

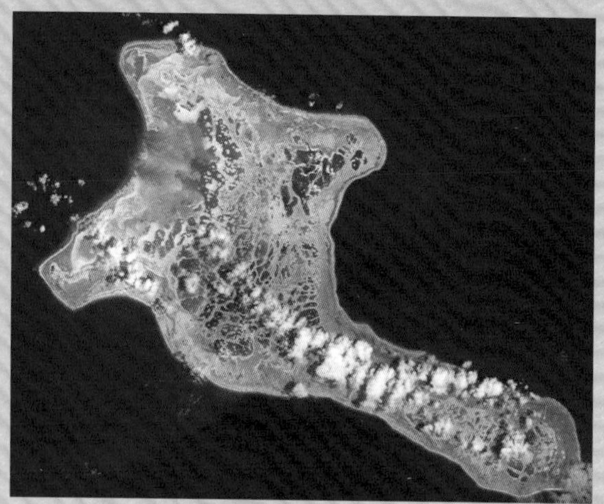

国际空间站第四探险队成员观测到的基里巴斯岛
Kiritimati Island as seen by the crew of Expedition 4 aboard the International Space Station

新奥尔良洪水，美国，2005年
The flood of New Orleans, USA, 2005

蓬皮杜中心广场，从巴黎圣圣梅里路看到的景象
Esplanade of the Center Georges-Pompidou, seen from rue Saint-Merri Paris

Carles Diaz、Xavier Sust、Oscar Tusquets和Lluís Clotet设计的兰布拉大街，巴塞罗那，1999年
Rambla del Raval by Carles Diaz, Xavier Sust, Oscar Tusquets and Lluís Clotet, Barcelona, 1999

Pelli Clark Pelli建筑师事务所设计的Gae Aulenti广场，米兰，2013年
Piazza Gae Aulenti by Pelli Clark Pelli Architects, Milan, 2013

泰晤士河水闸，伦敦，1984年
Thames Barrier, London, 1984

遭遇洪水的圣马可广场，威尼斯
Flooded Piazza San Marco, Venice

弗兰克·盖里设计的毕尔巴鄂古根海姆博物馆，1997年
Guggenheim Museum Bilbao by Frank Gehry, 1997

多米尼克·佩罗建筑事务所设计的法国国家图书馆，巴黎，1994年
National Library of France by Dominique Perrault Architecture, Paris, 1994

巴克敏斯特·富勒设计的蒙特利尔生态球，1967年
Montreal Biosphere by Buckminster Fuller, 1967

Bang Pa In漂浮馆，泰国
Bang Pa In Floating Pavilion, Thailand

De Urbanisten设计的Benthemplein广场，2013年
Water Square Benthemplein by De Urbanisten, Rotterdam, 2013

NLE建筑师事务所设计的马科科漂浮学校，拉各斯，2012年
Makoko Floating School by NLE Architects, Lagos, 2012

Michel Corajoud、Pierre Gangnet设计的水镜广场，波尔多，2006年
Water Mirror by Michel Corajoud, Pierre Gangnet, Bordeaux, 2006

该项目于 2003 年开始实施，但是两年前由于重大的腐败丑闻而中断，55 亿欧元预算中的 10 亿被用作政治回扣。如果"摩西工程"能完工的话，到时候，这一工程将会控制调节海平面上升 1m 以上可能发生的变化，但是这个海平面变化调节范围的预期到本世纪末可能已经不够应对海平面上升的速度了。似乎其他城市将支付不起像"摩西工程"或其仿效的模式——伦敦泰晤士河水闸（1984 年）这样昂贵的解决方案，因为它们都需要极高的运营和维护成本。此外，在许多国家，不仅仅在意大利，预算越大，腐败就越有可能发生。所以，或许早期的逐渐调整适应的方式更可取：如果你愿意冒险，不怕湿身，偶发的洪水也自有其乐趣。

鹿特丹市就是一个较好的例证，非常具有启发性。从城市建立之初到现在，鹿特丹市都处于海平面以下，海平面平均高出城市 2m。在这里，我们看到这个欧洲最大的港口非常重视海平面发生的巨大变化，对此认真设计规划。在过去的五年中，气候变化适应策略已经开始起作用。纽约市甚至聘请了鹿特丹三角洲城市工程的规划人员来预估这个伟大城市的水患难题。在众多的设计解决方案中，有一个方案建议把公共空间当作城市海绵。位于市区并距中央火车站不远的 Benthemplein 广场（De Urbanisten，2013 年）能够把雨水收集到三个不同的蓄水池中，用于引导洪水，并且最多能储存 36 个小时的暴雨雨量。这里的雨水通过渗透装置缓慢地渗入地面，而周边更大范围的雨水则被输送到开阔的水域。

鹿特丹的另外一个项目提出了未来公共空间也可以是浮动的这样的设计。2014 年，由公共领域建筑事务所设计的、位于鹿特丹 Kop van Zuid 区的浮动应对气候变化馆就建在浮动码头结构上。该建筑可以随着潮位上升和下降而上下起落 2m。根据巴

tional solution to flooding was to raise the embankments. Recently however, the wisdom of this tactic was challenged with the introduction of MOSE, a project for three gigantic adjustable dikes at the three mouths of the lagoon. Construction began in 2003 but was interrupted two years ago by major corruption scandals in which a billion euro of the 5.5 billion budget was siphoned off in political kickbacks. If and when completed MOSE will hold back a bit more than a one-meter change in sea levels, which will probably not be sufficient by the end of the century. It seems unlikely that other cities will be able to afford such expensive solutions as MOSE, or its model, the Thames Barrier in London (1984), both of which have exceptionally high operating and maintenance costs. Furthermore, in many countries, not just in Italy, the bigger the budget the more likely there will be corruption. So perhaps the earlier manner of slow adaptation is more desirable: temporary flooding has its joys if you are willing to risk getting a bit wet.

A better example of inspired adaptation, comes from Rotterdam, which since its origins has been on the average of 2 meters below sea level. Here we find the Europe's largest port seriously planning for the dramatic changes in sea levels. The Climate Adaptation Strategy during the past 5 years has begun to take effect. New York City has indeed hired planners from the Rotterdam Delta Cities program to anticipate that great city's aquatic dilemma. Among the design solutions, one design firm proposes public space as an urban sponge. The Benthemplein Square (De Urbanisten, 2013), set in an urban quarter not far from the central station, has been designed to catch run-off water in three different basins, steering the flood and delaying the storm water for a maximum of 36 hours. The local water is slowly released into the ground via infiltration and the water from

克敏斯特·富勒的网格穹顶结构原理，泡泡状的展馆采用高科技手段来实现其目的，建成了高达 25 层的浮动建筑。每一个泰国人都知道，低技方法也是可行的。三年前在尼日利亚首都拉各斯建的马科科漂浮学校（孔勒·阿德耶米等人设计）最近进行了重建并在威尼斯双年展上亮相，马科科漂浮学校为易受洪水影响的社区提供了一个既廉价又优质的公共场所。

无论是漂浮在水上还是浸透在水中，全球变暖时代的公共空间既需要考虑生态环保，又需要为人们带来一种场所感。虽然波尔多水镜广场（Michel Corajoud、Pierre Gangnet 和水力工程师 Jean-Marc Llorca，2006 年）最初设计时并没有想要表现水位的波动，但实际上却做到了这一点，成为欧洲最受欢迎的公共空间之一。该广场沿着新电车轨道线延伸，面积超过 3000m²，铺着光滑的花岗岩，当水浸满广场表面时，宁静的水面会倒映建于 18 世纪的证券交易所。白天，水通过裂缝汩汩地流出，薄雾升腾，然后水又神秘地渐渐消失，就像潮汐的变化一样。水镜广场所传达的是一种乐观的心态，即使水将对未来的城市空间构成威胁，但是也能成为美丽和与人们快乐互动的源泉，人们沉浸其中，可能把手机也抛在脑后。

wider surroundings is transported to a body of open water.
Another project in Rotterdam proposes that the public space of the future can also be floating. The Floating Climate Proof Pavilion, moored at Rotterdam's Kop van Zuid district, was constructed by Public Domain Architecten on a floating pier in 2014. It rises and falls with the two-meter change in tide levels. Based on Buckminster Fuller's geodesic dome structures, the bubble-shaped pavilion uses high-tech methods to achieve its goals, promoting 25-story buildings that float. Lower tech methods, are just as possible, as anyone in Thailand knows, and 3 years ago the Makoko Floating School in Lagos, Nigeria (Kunlé Adeyemi, et al.), recently reconstructed at the Venice Biennale, provided a cheap and excellent public venue in the midst of a flood prone community.
Whether floating or soaking, public space in the age of global warming needs to perform ecologically while offering a sense of place. While not intended to symbolize the volatility of water levels, the Water Mirror in Bordeaux (Michel Corajoud, Pierre Gangnet, and hydraulic engineer Jean-Marc Llorca, 2006), does just that and has become one of the most popular public spaces in Europe. It stretches more than 3,000 square meters along the new tram line on a smooth granite plane that when drenched in calm water reflects the 18th-century facades of Place de la Bourse. During the day water gurgles up through the cracks, mist pours out, and water mysteriously drains away much like the rhythm of the tides. The Water Mirror transmits an optimistic sentiment that even if water will be a threat to the urban space of the future, it can also become a source of great beauty and joyful interaction, so much so that people may forget to use their cell phones. Richard Ingersoll

雅高酒店竞技场馆

DVVD

HOTELS ARENA
RTL

法国最大的音乐会、表演和体育赛事举办场地巴黎贝尔西综合体育馆（POPB）于 1984 年 2 月开始投入使用，从此以后，其强大的多功能赛事承办能力使其成为一个独特的举办全球活动的场所。这个由蓝色金属网状结构、斜坡式草坪和大胆的金字塔体量组成的建筑独树一帜，在巴黎的城市景观中留下了浓墨重彩的一笔，也给因各类赛事、演唱会而到来的参观者留下了极其深刻的印象。不过，从快速帆板到赛车，从麦当娜演唱会到特技滑雪，经过三十余年的高频使用，该场馆需要进行一次全面的改造来增强其承办能力，创造一个人们如今所需要的舒适的环境。2011 年，DVVD 建筑事务所被选中，对这个无论是技术还是功能方面都野心勃勃的项目进行升级改造。事务所本着尊重原建筑的原则进行改造，使改造后的雅高酒店竞技场馆成为巴黎冲击 2024 年夏季奥运会的又一助力。

DVVD 事务所面临的主要挑战就是改变贝尔西综合体育馆固有的空间条件，老建筑是一座与周边环境脱节、道路系统指示不明、不为公众着想的内向型建筑。为了解决上述问题，建筑师们拆除了似乎永无止境的花岗岩台阶和面向贝尔西大街的高层停车场，取而代之的是一个 2500m^2 的广场。这一大厅式的空间填补了音乐厅与街道之间的空白，成为面向城市开放的通道。而且，它改善了原建筑结构，既与公众连为一体，又为公众提供了庇护空间。同时，设计巧妙的堆叠储物系统使音乐厅观演平台的可折叠座椅得以就地存放，原来存放座椅的空间被改造成拥有 90 个停车位的停车场。其他对基础设施的改造还包括塞纳河畔的玻璃天棚，公园一侧的人行道以及新设的 VIP、溜冰场和新闻媒体空间的入口。这些改造与该建筑独特的金属网状结构和斜坡式草坪融为一体，使其从巴黎天际线中脱颖而出。

现在，场馆在面朝贝尔西大街和塞纳河畔两侧都有入口，仿佛在邀请公众进入建筑内部参观。由木材、钢结构和玻璃打造的新接待空间就像原来就存在的草坪和混凝土之间的分界线一样，为人们提供了会面与交谈的空间，人们在这里可以清晰地看到音乐厅。接待空间向公众全天候开放，真正地将建筑变为公众日常生活的一部分。这里生机勃勃，酒吧、餐厅、商店等各种辅助功能区吸引了无数的回头客。玻璃天棚使整个接待空间内部沐浴在阳光中。玻璃天棚不仅重新诠释了金属网状结构，并且给视野中的天空和屋顶草坪定义了边框。具有动感且精致的木条勾勒出斜坡式屋顶的轮廓，并一直延伸到楼梯和扶梯处，把原本毫无魅力又不便捷的步行长廊改造成富有时代感的地方。

面向大街一侧的公共广场微微后退于高速公路。作为该建筑正面的公共空间，它将对周边地区及其居民产生重大影响。另外，缓缓倾斜的屋顶平台对所有公众开放，成为户外休闲的好去处，人们可以在阳光下放松、玩耍。而且，建筑师为残疾人设计了专用通道，另有一座步行天桥将建筑与附近的贝尔西公园和市图书馆连为一体，使雅高酒店竞技场馆成为一个羽翼丰满的城市枢纽。

DVVD 事务所的建筑师们将新鲜愉悦的生活注入到这个传奇的巴黎场馆。经过巧妙而一丝不苟的改造，雅高酒店竞技场馆已跻身世界最重要的多功能体育场馆之列，可承办高质量赛事和演出、提供热情周到的服务，且具有强大的功能和设施。它完美地回应了 21 世纪对场馆的最新要求，宏伟壮观，引人入胜，重新回归到巴黎核心区域的中心舞台。

Accorhotels Arena

The largest venue for concerts, shows and sports events in France, the Palais Omnisports de Paris-Bercy (POPB) has become a unique global venue with multi-purpose capability, since it opened in February 1984. Its blue metallic web structure, sloping lawns and bold pyramidal form has made a strong impression upon the urban landscape and captured the imagination of the public by performances of leading singers and athletic games of sports stars. However, after three decades of intensive use, from funboarding to stock car racing, and from Madonna concerts to acrobatic ski-ing, the arena was in need of renovation to be able to host events and provide the comfort required of today. For the upgrade project, ambitious both in technical and functional aspects, DVVD was selected to do the job in 2011. Their intervention

respecting the existing structure, delivers fresh impetus and new prospects for the coming 2024 summer Olympics. The main challenge for DVVD was to change the spatial conditions of the old POPB; an inward-facing facility, substantially disconnected to its surroundings, with confusing access routes and little consideration for the public. For this, the architects created a 2,500m² concourse replacing the interminable granite stairways and high-level car parks toward the nearby Rue de Bercy. Bridging the gap between the concert hall and street, this lobby space now provides an opening onto the city. Also, reinforcing the existing structure, it serves as a link and shelter to the public. Meanwhile, the parking lots for 90 vehicles were relocated to the former storage spaces, where the once retractable terraced seats are now stored in the concert hall, thanks to an ingenious system of stacking drawers. Other renovations to the pedestal structure include

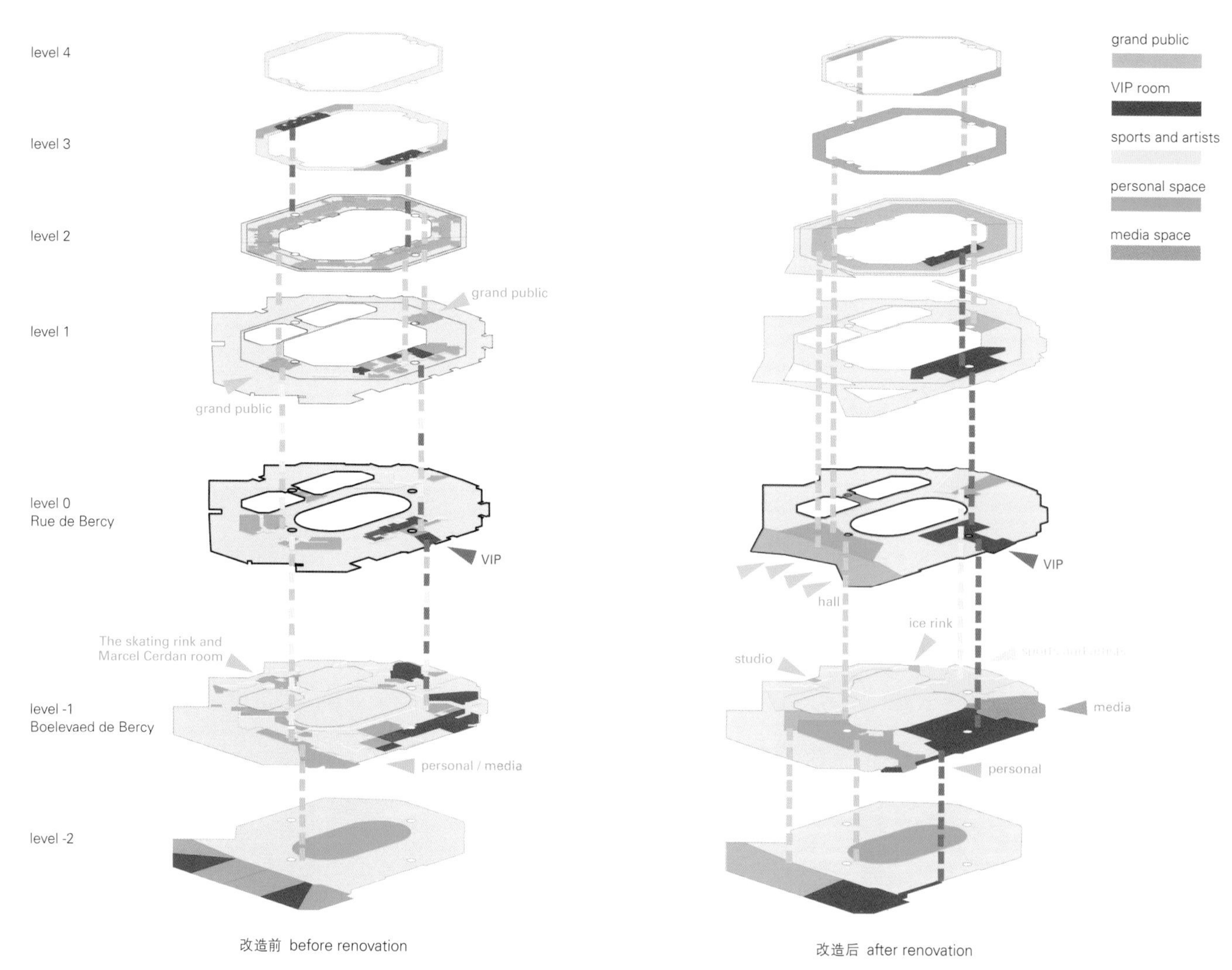

改造前 before renovation

改造后 after renovation

田径运动 athletics

冰上运动 ice sports

帆板运动 sailboard

篮球运动 basketball

拳击运动 boxing

音乐厅前台
concert hall front stage

音乐厅横向舞台（半场）
concert hall transverse stage (semi-arena)

音乐厅中心舞台
concert hall central stage

PORTES de U à K

PORTES de U à K

ACCORHOTELS ARENA

the glass canopy on the River Seine side, the walkway on the park side, and new entrances for the VIP, ice rink, and press. These changes are incorporated into the building's strong identity comprising of metal web-work and sloping lawns, a distinctive part of the Parisian skyline.

The arena now opened towards the Rue de Bercy and the Seine, consequently invites the public into its interior. New reception areas, constructed of timber, structural steelwork, and glass, resemble the pre-existing boundary of lawn and concrete, and provide spaces for meeting and conversation, while giving a clear view to the concert hall. These spaces are freely accessible throughout the day, genuinely transforming the venue into a space for the everyday life. The subtly vivid atmosphere through various supportive functions such as bars, restaurants, and shops asks for a return visit to its users. The interior of the reception space, is bathed in light, thanks to the glazed canopies which both reinterpret the metal web structure and frame astonishing views of the sky and roof lawns. Here, the roof slope is outlined and reinforced by dynamic and sophisticated timber laths, which also extends into the stairways and escalators, transforming a previously charmless and inconvenient pedestrian walkway into a place

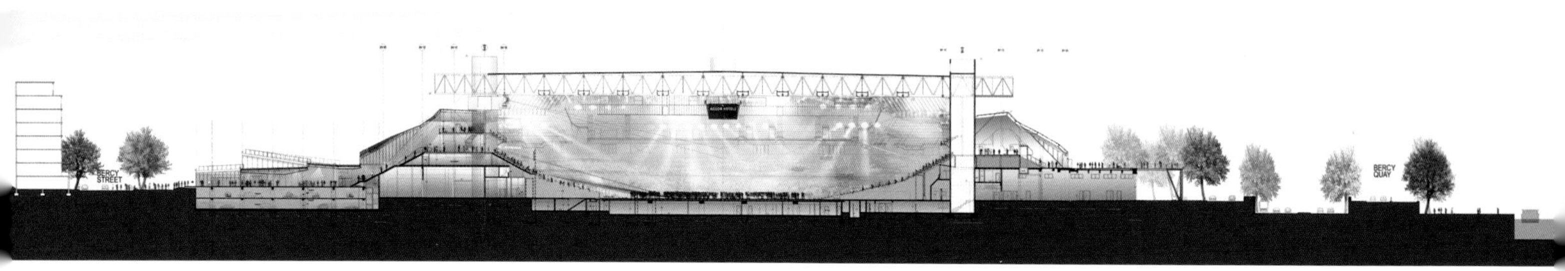

B–B' 剖面图 section B-B'

esquisse fauteuil grand public.

- Les mêmes principes sont valables pour les places à proxiation
à différence que l'assise est complètement rembourrée

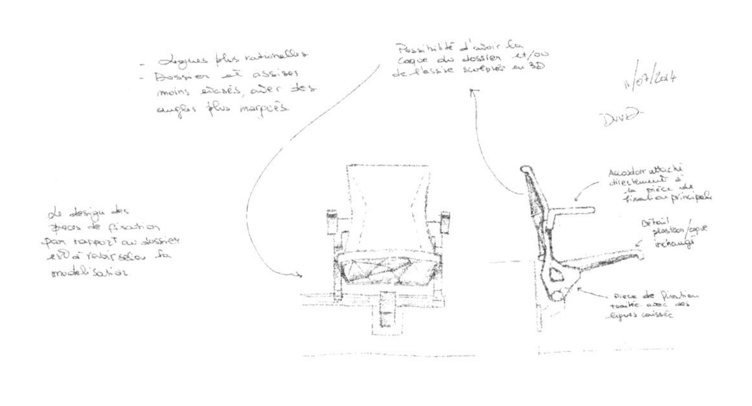

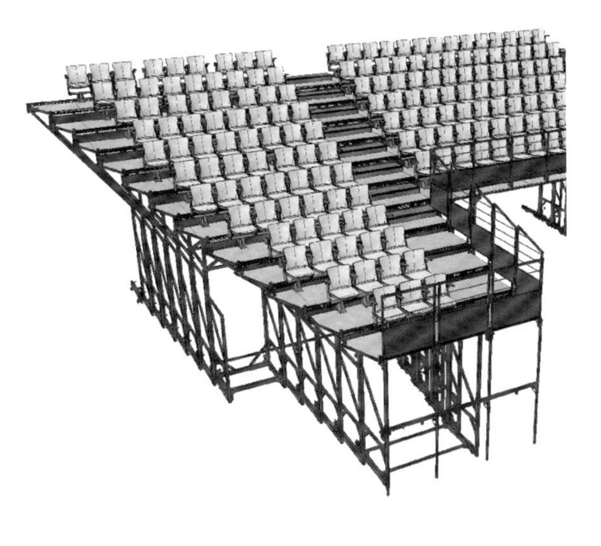

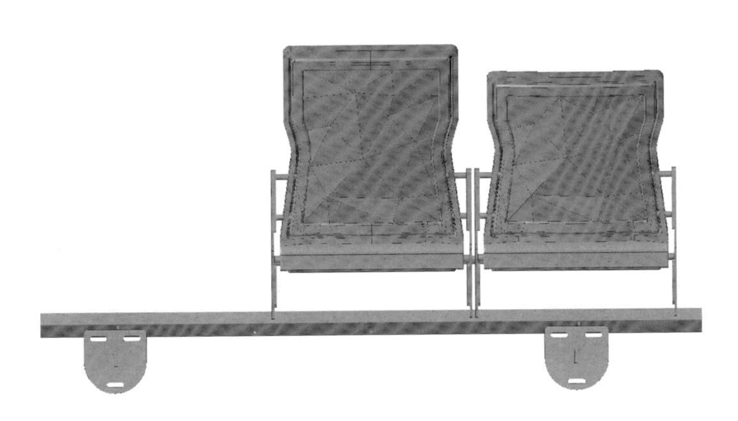

reflecting the passing of time.

The concourse facing the street, is slightly set back from the highway. This frontal public space will fundamentally influence the nearby neighborhood and its inhabitants. Along with this, the gently inclined roof, accessible to all, will become a favorite outdoor spot to relax and play in the sunshine. Also, incorporating a ramp for the disabled and an overpass that connects with the adjacent Parc de Bercy and municipal library, the Accorhotels Arena will become a fully-fledged urban interchange.

The architects have breathed new and joyful life into this legendary Parisian venue. Thanks to an ingenious and meticulous renovation, the Accorhotels Arena now ranks third among the leading multi-purpose sports arenas of the world, with high quality performances, hospitality, functional capabilities and facilities. By responding to the latest of the 21st century, it has spectacularly returned to center stage in the heart of the capital city.

项目名称：Accorhotels Arena / 地点：Paris (12th arrondissement), France / 建筑师：DVVD / 项目团队：Paula Castro, Céline Cerisier, Vincent Dominguez, Toma Dryjski, Bertrand Potel, Louis Ratajczak, Daniel Vaniche / 项目主管：Vincent Dominguez and Daniel Vaniche / 项目经理：Fulvia Parlati, Louis Ratajczak, Monica Sierra
代理项目经理：M. Dominguez, M. de Feo, N. Didier, B. Frati, E. Glass, J.Chelza, A. Hery, C. Lapassat, R. Pericaud, L. Piciocchi, A. Rivera, C. Walsh / 总承包商：Bouygues Bâtiment Île-de-France / 设计公司：Alto Ingénierie_fluids, HQE, fire safety systems, Casso_technical assistance for the coordination of fire safety systems, Cronos conseil_safety and risk prevention, Peutz_acoustics, QS Cube_economics, Sepia GC_geotechnical and civil engineering, Systal_kitchens / 改造前楼面面积：59,100m² / 改造后楼面面积：62,000m² / 预算：110 million euros(30% for pahse 1 and 70% for phase 2) / 总工期：17 months / 竣工时间：2015.10
摄影师：©Sergio Grazia (courtesy of the architect)

DÉLI
DÉLI

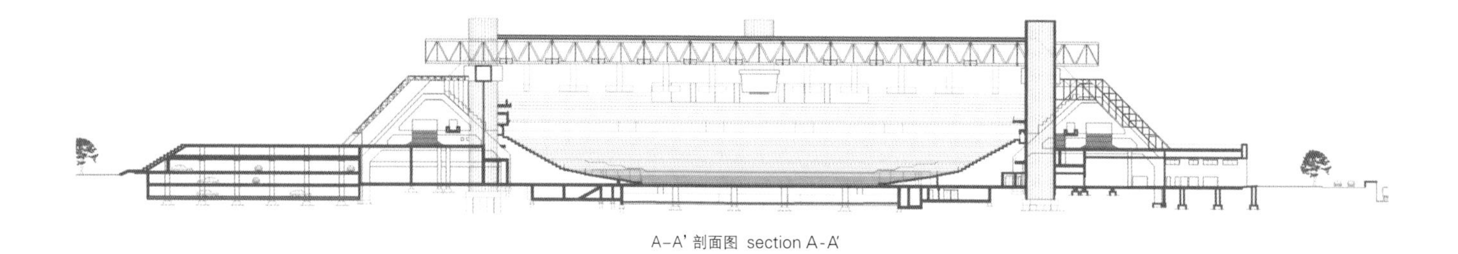

A–A' 剖面图 section A-A'

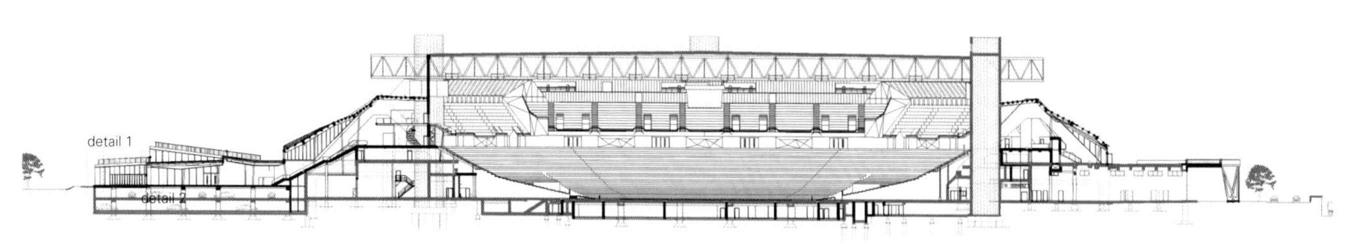

B–B' 剖面图 section B-B'

1. roofing complex
   - wood deck
   - concrete screed
   - waterproofing
   - insulation
2. structure
   - composite steel and concrete slab
   - steel girders
3. false ceiling
   - insulation
   - wood lattice cladding (suspended)
4. steel protection gate
5. insulated cladding panel with aluminum finishes
6. glass door with aluminum frame
7. steel grating
8. balustrade
   - metallic handrail including lighting device
   - steel mesh
9. powder-coated steel mullion (facade structure)

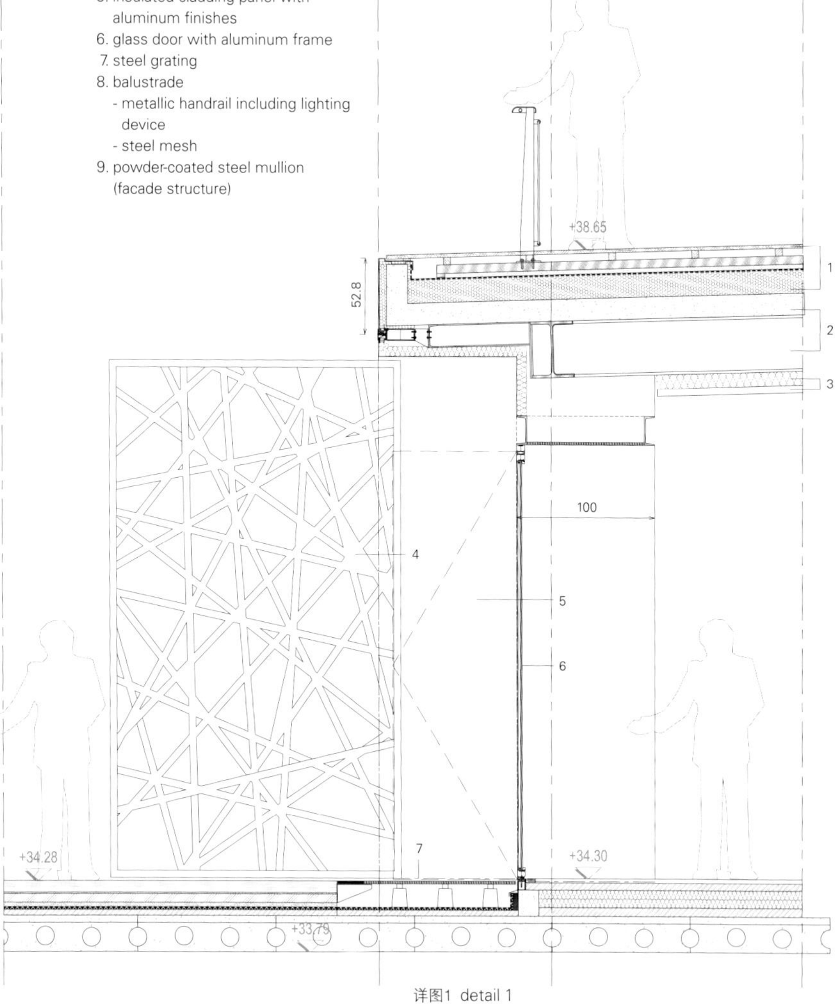

详图1 detail 1

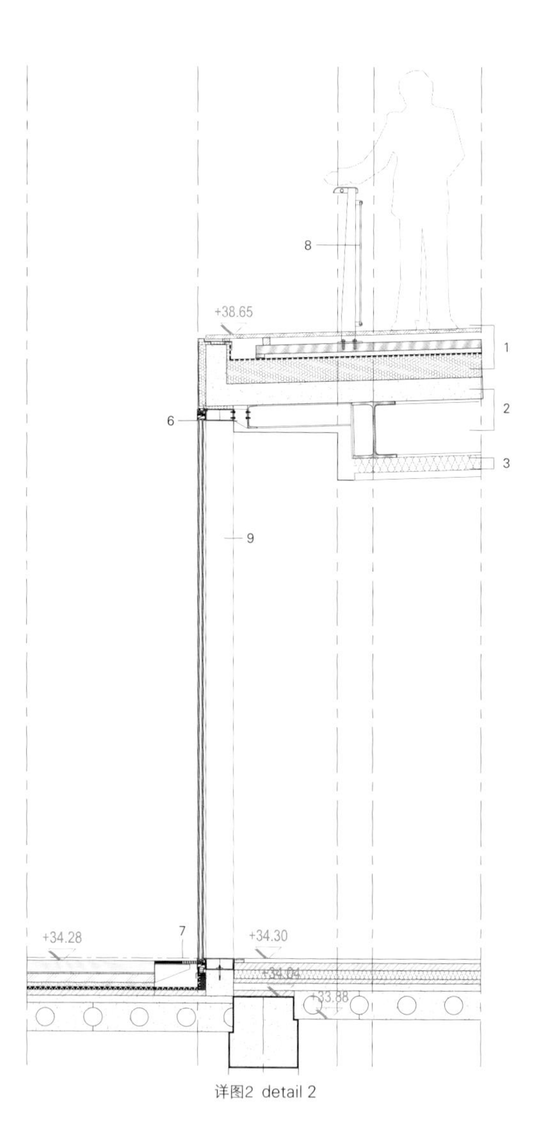

详图2 detail 2

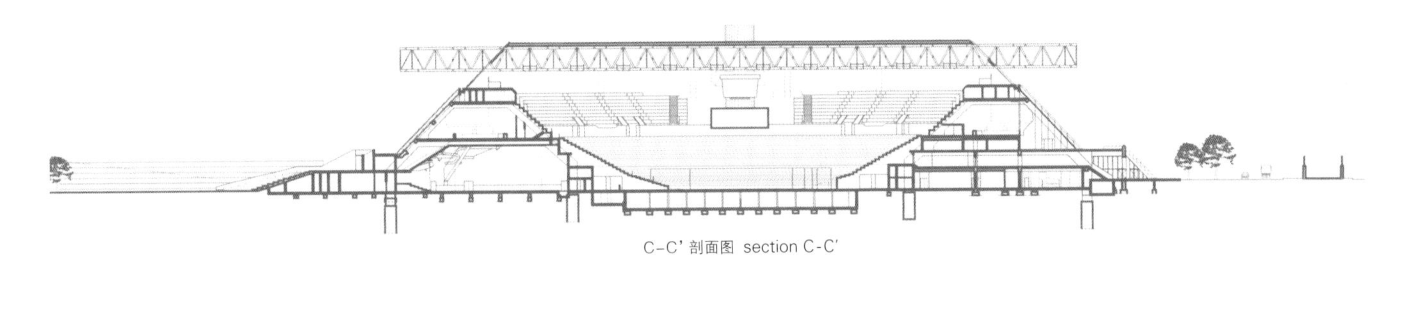

C-C' 剖面图 section C-C'

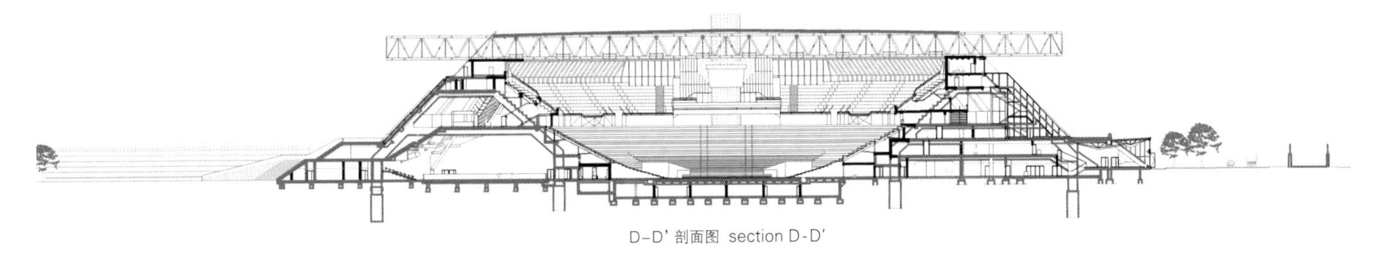

D-D' 剖面图 section D-D'

斯特拉斯堡会议中心

Dietrich | Untertrifaller Architects + Rey-Lucquet et associés

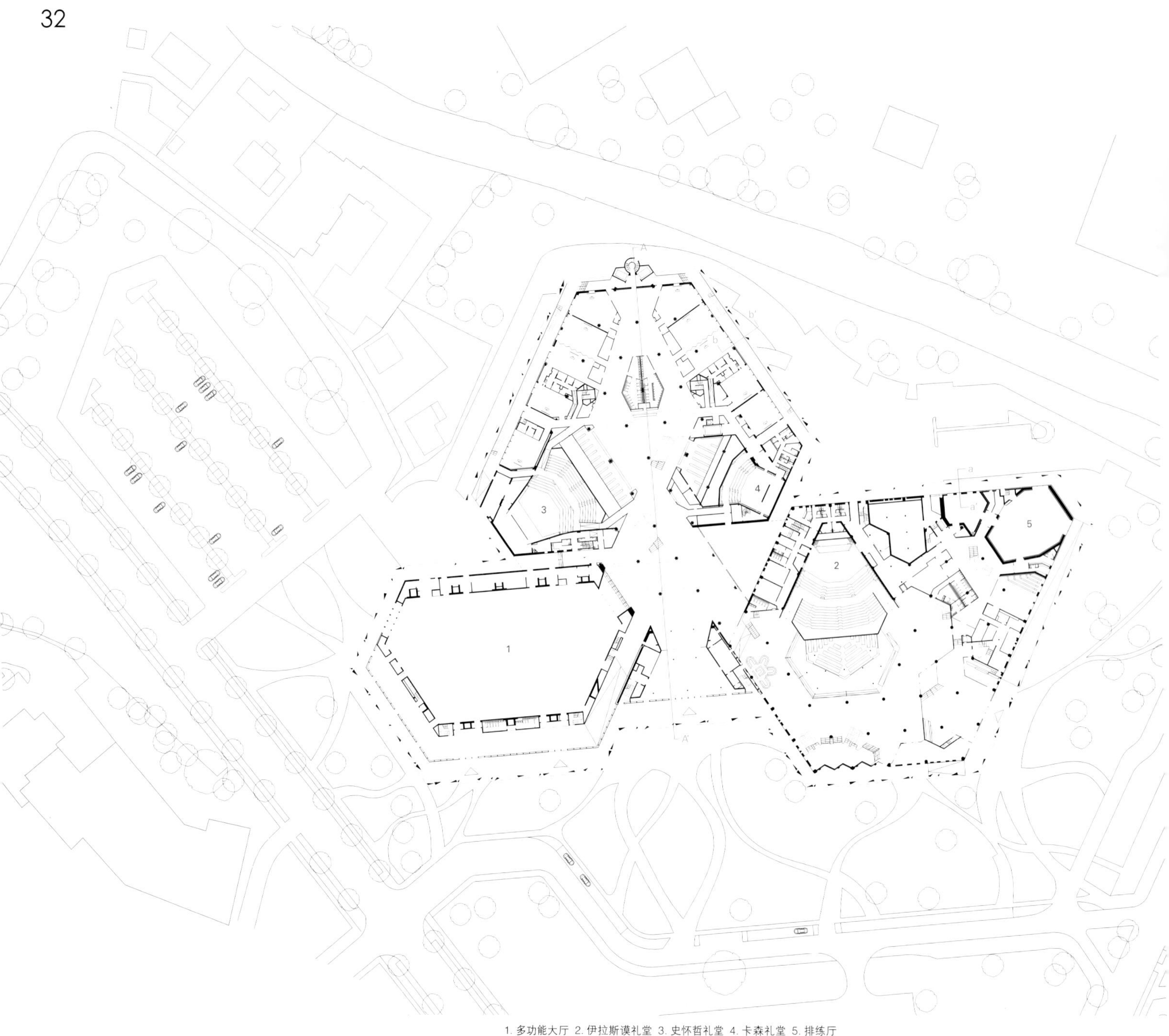

1. 多功能大厅 2. 伊拉斯谟礼堂 3. 史怀哲礼堂 4. 卡森礼堂 5. 排练厅
1. multi-functional hall 2. auditorium Erasme 3. auditorium Schweitzer 4. auditorium Cassin 5. rehearsal hall
一层 first floor

斯特拉斯堡音乐厅和会议中心的设计由奥地利建筑公司 Dietrich | Untertrifaller 和法国公司 Rey-Lucquet et associés 联合设计完成。该项目将两座分别建于 20 世纪 70 年代和 20 世纪 80 年代的老音乐厅与新建筑结合在一起，从而创建了一个建筑形象独特而又和谐的整体空间。会议中心的扩建与常规改造项目包括：建造一个 3000m^2 的多功能厅、一个 450 席的会议厅和一个 520 席的礼堂；扩建与改造两个原有的音乐厅；为斯特拉斯堡爱乐乐团新建一个排练厅。

为了实现设计，Dietrich | Untertrifaller 和 Rey-Lucquet 两家公司决定在原来的等边三角形核心形状基础上进一步开发和扩建。同时，新建筑和不锈钢拱廊立面为整个会议中心勾画出一个全新的轮廓。建筑师们将尽可能多的有用功能整合到原有建筑中，并在新建筑中继续保持典型的六边形的布局特点。集宽敞明亮的空间、画廊和纽带作用于一体的中央大厅连通着音乐厅、会议室和展览空间。这种清晰的布局及开放的空间设计简化并构建了整个综合体的功能流程分布，给予访客更好的方向感。建筑外立面的设计使整个设计理念一目了然：近 1km 长的拱廊环绕包围着整个建筑群，使其外形新颖独特。每根 15m 高、6t 重的钢柱表面由角度、形状不一的不锈钢钢板饰面，形成一种带有扭转韵律感的动感外壳，格外引人注目。

史怀哲礼堂可容纳的人数从 900 人扩大到了 1200 人，有 1900 个座位的伊拉斯谟礼堂经过优化后，可以举办音乐会和会议。位于这两座建筑西面的是新建的六边形多功能厅，它与原来的建筑融为一体，并与老建筑一起共同构成了一个新的主入口和前厅。新扩建的部分包括更人性化的蓝色蒙克大厅以及位于入口大厅上层、可以鸟瞰整个中央大厅和园区的亮红色玛丽·库里活动大厅。与伊拉斯谟礼堂紧邻的是新建的斯特拉斯堡爱乐乐团排练厅，排练厅用橡木饰面装饰。斯特拉斯堡音乐厅和会议中心现在包括一个大型多功能厅、三个礼堂、两个会议厅、15 间会议室、休息室、酒吧、一个餐厅，还有办公区域和一个停车场。

斯特拉斯堡音乐厅和会议中心总楼面面积为 44 500m^2，是瓦肯－欧洲城市发展项目的一个重要部分，使这个位于城市中心北部、靠近欧

0 25 50m

洲议会的区域拥有了一个新的建筑地标，成为一个国际性的商务与服务地点。

Strasbourg Convention Center

The design for the Strasbourg Palais de la Musique et des Congrès by Austrian architectural firm Dietrich | Untertrifaller Architects and French firm Rey-Lucquet et associés combines the two existing music halls from the 1970s and 80s with new buildings to create a harmonious ensemble with a distinctive architectural identity. The expansion and general renovation of the convention center involves the construction of a 3,000 m² multi-functional hall, a conference hall for 450 people, and a 520-seat auditorium, the expansion and conversion of two existing concert halls, plus a new rehearsal hall for the Strasbourg Philharmonic orchestra.

For their design, Dietrich | Untertrifaller and Rey-Lucquet decided to further develop and expand on the existing central motif of equilateral triangles. They also created a completely new silhouette with the new buildings and stainless steel arcades. The architects integrated as many useful functions as possible into the existing buildings and continued the use of the typical hexagon in the new buildings. A central foyer with air spaces, galleries and bridges unites and connects the concert, conference and exhibition spaces. This clearly arranged and open spatial design simplifies and structures the complex's functional processes and provides visitors with

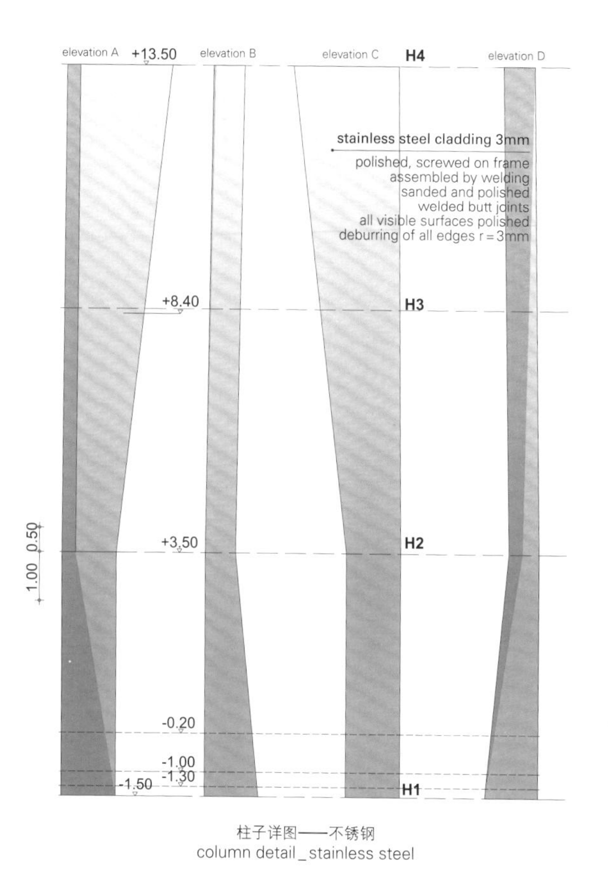

柱子详图——不锈钢
column detail_stainless steel

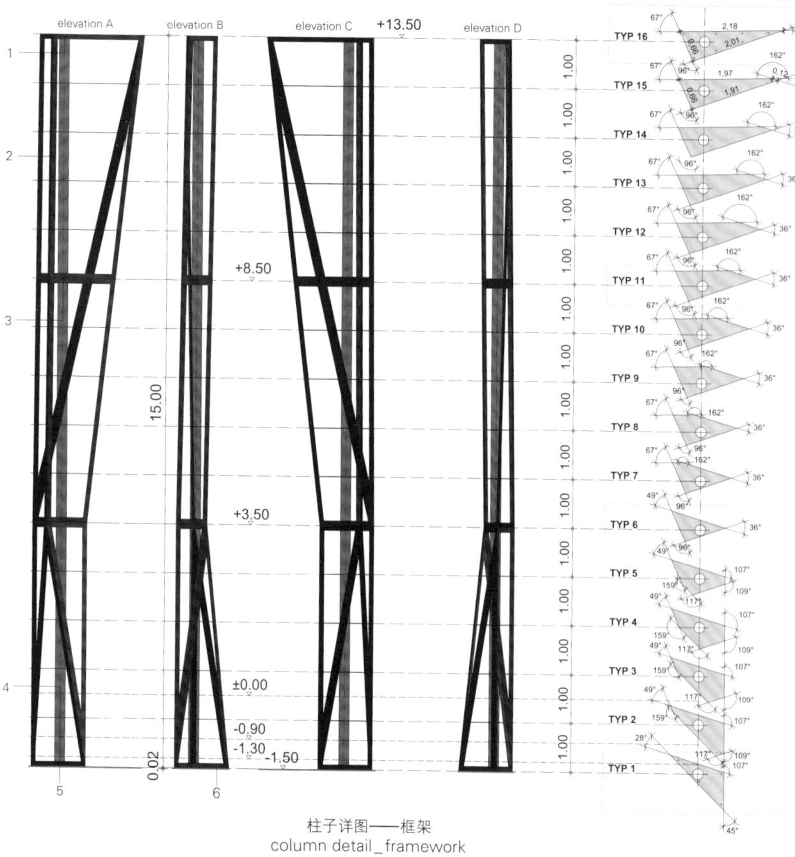

柱子详图——框架
column detail_framework

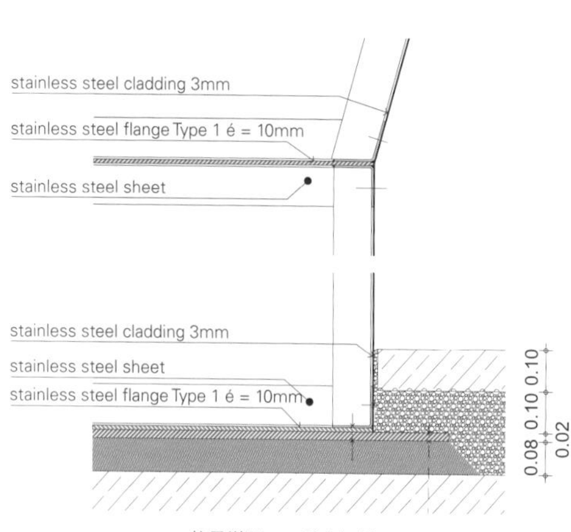

柱子详图——垂直切割
column detail_vertical cutting

1. **steel sheet stainless steel**
   type 4 é = 10mm
   assembling openings,
   punctual reservations
   to be provided for
   installation of the insulation
2. **stainless steel frame of angles**
   (stainless steel)
   100x100x4mm
   alpha individual assembling
   openings for the outer
   covering made of stainless
   steel
3. **stainless steel sheet**
   (constructive reinforcement)
   Type 3 é = 10mm
   assembling openings,
   punctual reservations to be
   provided for installation
   of the insulation of the
   cavity
4. **stainless steel flange**
   Type 1 é = 10mm
   assembling openings,
   punctual reservations to be
   provided for installation of
   the insulation of the cavity
5. **galvanized base plate**
   1200x é = 20mm
   assembling openings,
   reservations for drainage
   required
6. door drape mineral panel

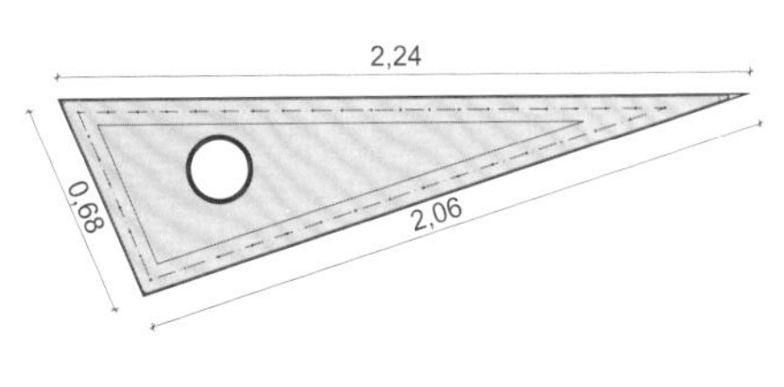

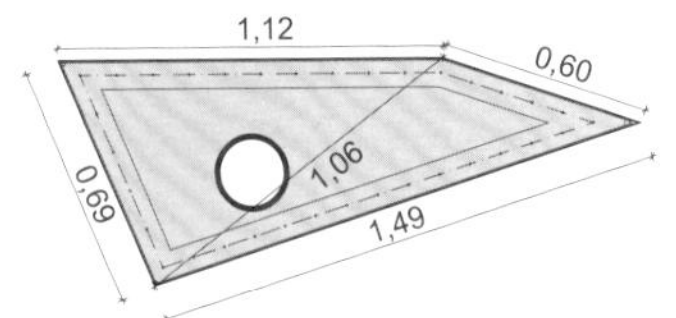

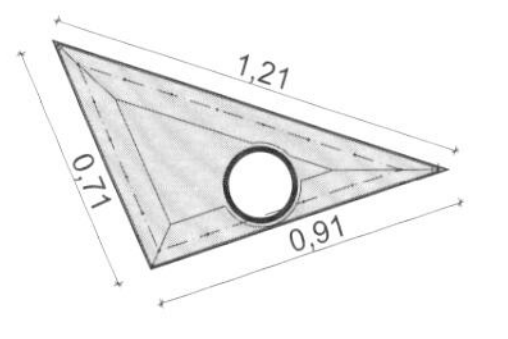

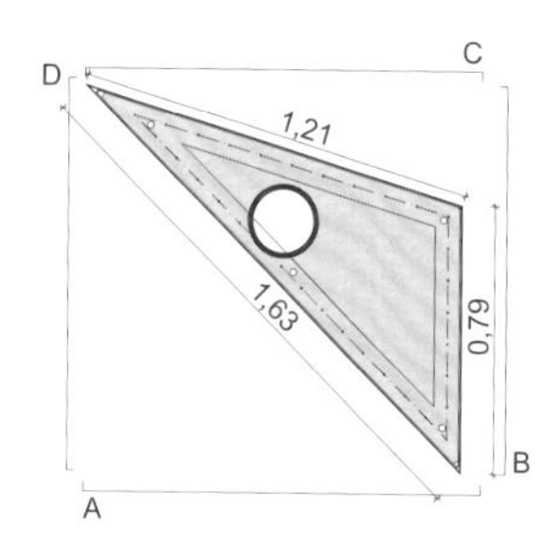

柱子详图——不锈钢凸缘
column detail_stainless steel flanges

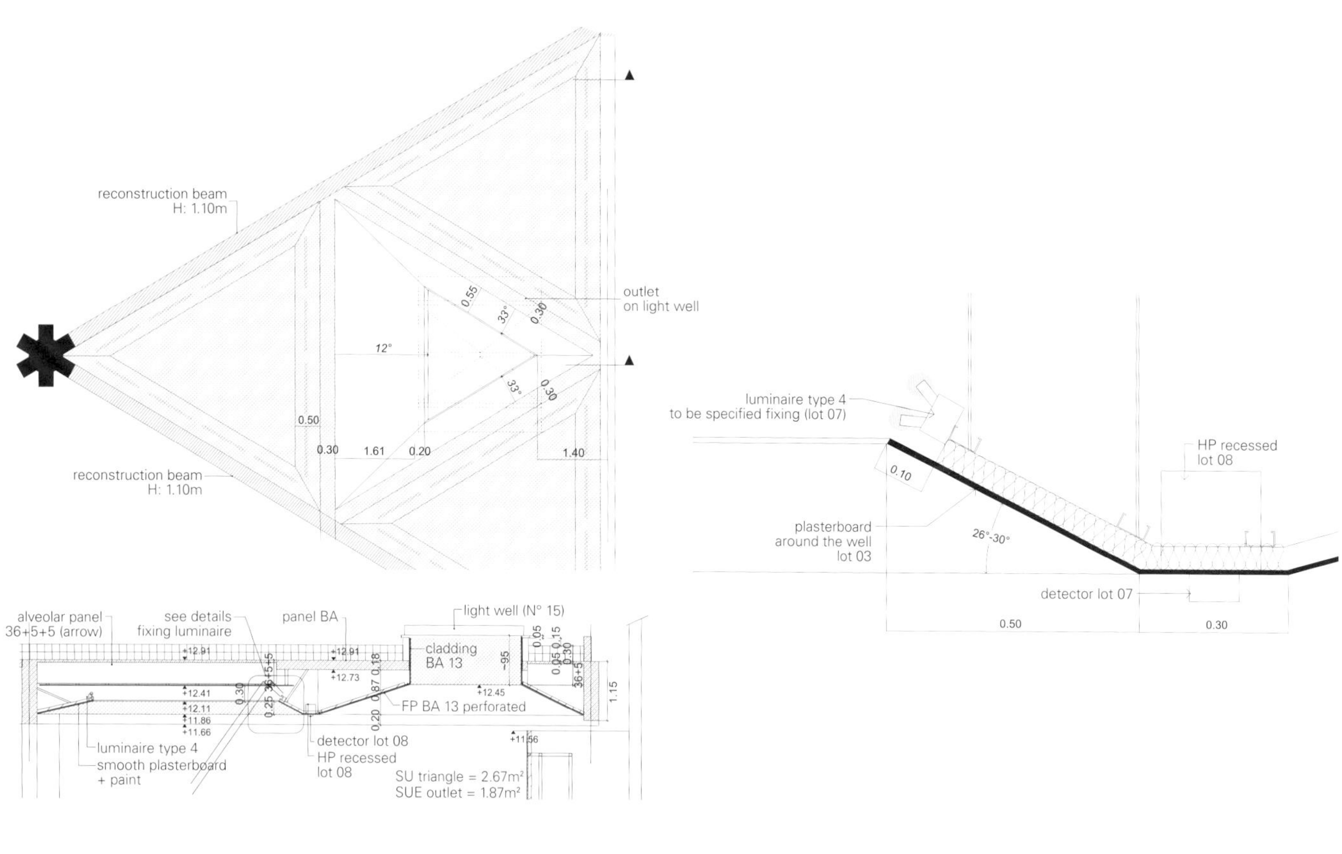
reconstruction beam
H: 1.10m
outlet
on light well
0.55
33°
0.30
12°
33°
0.30
0.50
0.30
1.61
0.20
1.40
reconstruction beam
H: 1.10m
luminaire type 4
to be specified fixing (lot 07)
HP recessed
lot 08
0.10
plasterboard
around the well
lot 03
26°-30°
detector lot 07
0.50
0.30
alveolar panel
36+5+5 (arrow)
see details
fixing luminaire
panel BA
light well (N° 15)
+12.91
+12.91
+12.73
cladding
BA 13
-95
+12.41
+12.11
+11.86
+11.66
+12.45
FP BA 13 perforated
1.15
+11.56
luminaire type 4
smooth plasterboard
+ paint
detector lot 08
HP recessed
lot 08
SU triangle = 2.67m²
SUE outlet = 1.87m²

improved orientation. The architectural concept is also visible on the facade: An almost kilometer-long arcade wraps around and encloses the entire building complex, giving it a distinct profile. The 15m high and 6 ton steel columns covered in angularly folded stainless steel sheets form an attractive, dynamic outer shell with their twisting rhythms.

The capacity of the Auditorium Schweitzer was expanded from 900 to 1,200 visitors and the 1,900-seat Auditorium Erasme was optimized for concerts and conferences. Located just to the west of these two buildings is the new hexagonal multifunctional hall. It is integrated into the existing structures and together with the old buildings forms the new main entrance and forecourt. New additions include the blue Munch hall with its more intimate atmosphere and the Marie Curie event hall in radiant red floating above the entrance hall with a view of the foyer and the park. Built in close proximity to the Auditorium Erasme the Strasbourg Philharmonic orchestra received a new oak-clad rehearsal hall. The Palais de la Musique et des Congrès now comprises a huge multifunctional hall, three auditoriums, two conference halls, fifteen conference rooms, foyers, bars, a restaurant, office spaces and a parking garage.

With a gross floor area of 44,500m², the Palais de la Musique et des Congrès is an important part of the Wacken-Europe urban development project. The district located in the north of the city center close to the European Parliament receives a new architectural landmark and positions itself as an attractive international business and service destination.

项目名称：Strasbourg Convention Center / 地点：F-67000 Straßburg, Place Bordeaux, France
建筑师：Dietrich | Untertrifaller Architects & Rey-Lucquet et associés / 项目管理：Heiner Walker (D | U) & Aurélien Vollmar (R-L)
结构/建筑工程：OTE Ingénierie, Illkirch / Serue Ingénierie / 可持续性&建筑物理：Solares Bauen
成本规划：C2BI / 声学设计：Müller-BBM / 舞台规划：Walter Kottke
立面规划：CEEF / 景观规划：Digitale Paysage / 厨房规划：Ecotral / 客户：Eurométropole de Strasbourg
有效楼层面积：44,500m² / 可容纳人数：ca. 15,000 visitors / 竣工时间：2011 / 施工时间：2013—2016
摄影师：©Bruno Klomfar (courtesy of the architect)

1. 玛丽·库里活动大厅 2. 餐厅 1. Marie Curie event hall 2. restaurant
A-A' 剖面图 section A-A'

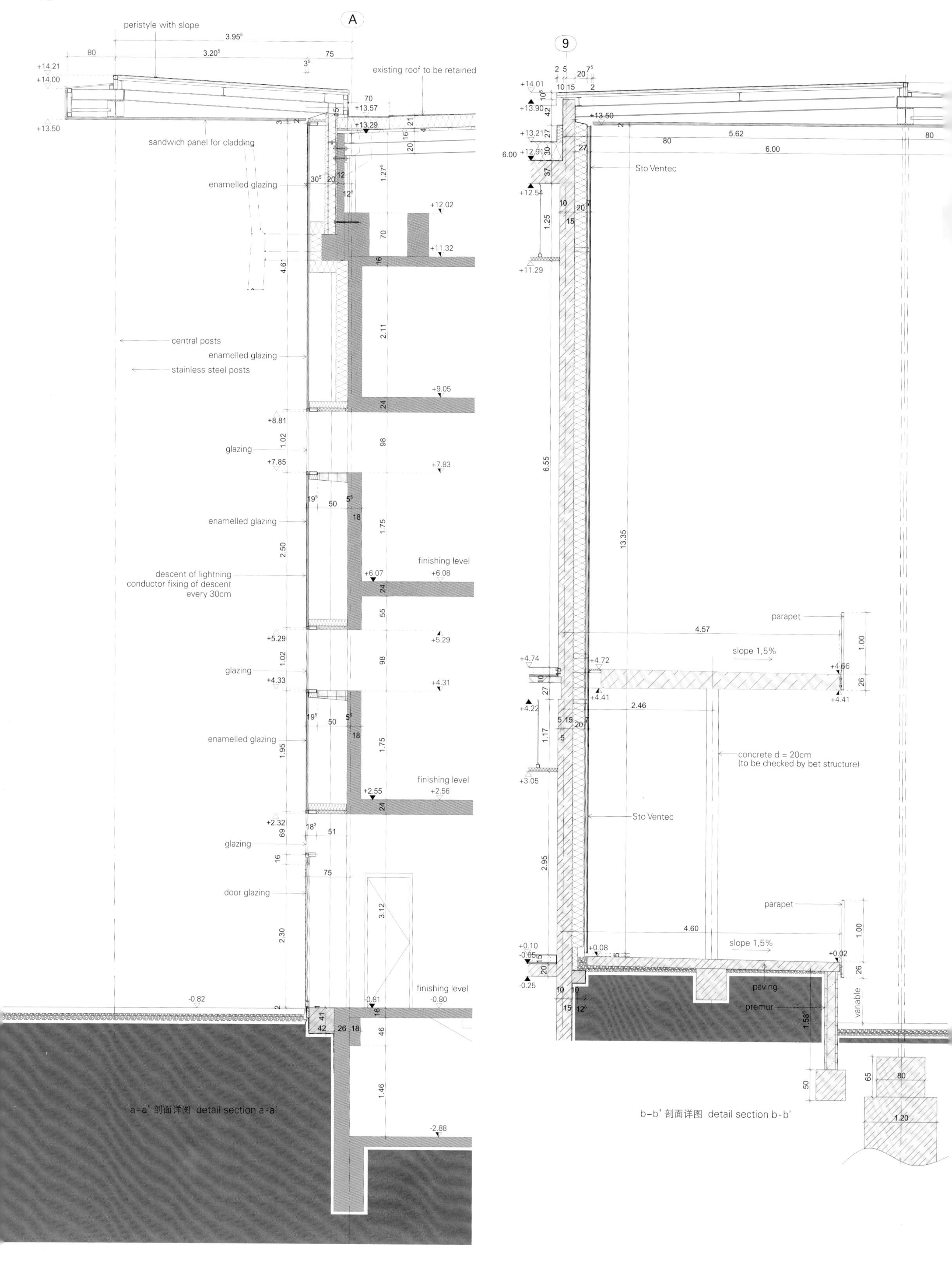

a–a' 剖面详图 detail section a-a'

b–b' 剖面详图 detail section b-b'

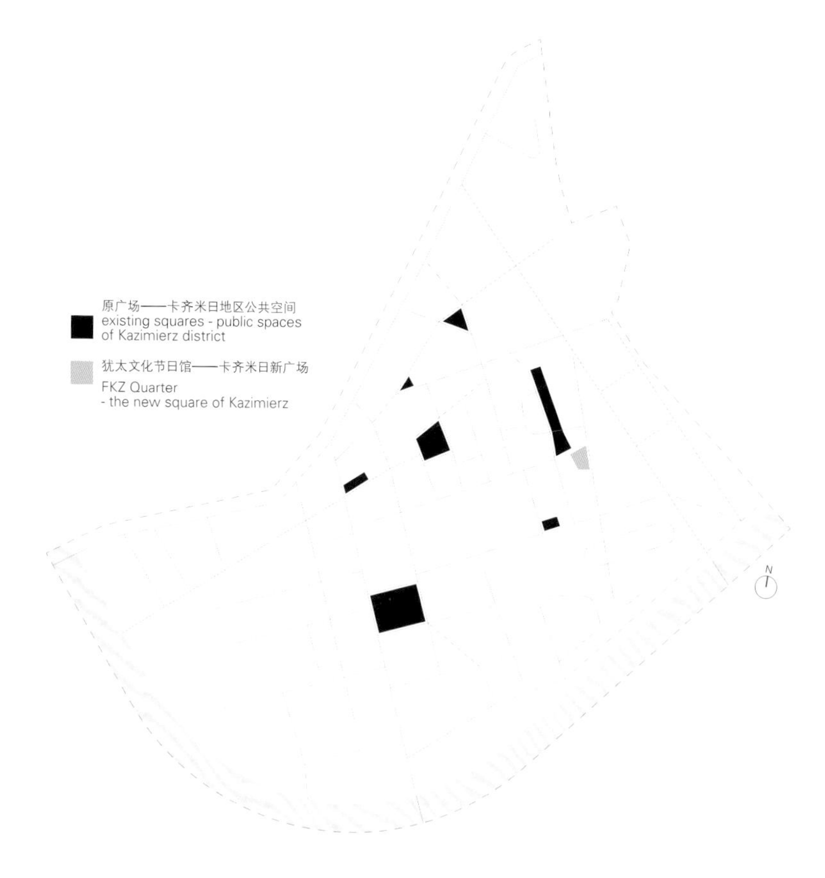

犹太文化节日馆

BudCud

2015 年，第 25 届克拉科夫犹太文化节在克拉科夫犹太人社区卡齐米日以及世界其他犹太地区同时举办。根据这一主题，BudCud 事务所在这个区域几处空旷广场中的一处，设计了一个临时公共空间。建筑师注意到卡齐米日的广场最能体现节日意义的特征，可惜的是，广场都被汽车、古迹或者商业活动占满。随着节日的到来，建筑师们决定搭建一个犹太文化节日馆，一个能将社会生活融入其中的建筑设施。这个临时公共城市空间由多个不同功能模块组成，围绕一个稍稍抬高的中央广场布置。

犹太文化节日馆位于老犹太教堂后身人迹罕至的 Dajwor 街上，但是这条街上经常停满了旅游大巴。项目所在地是一处带有自然坡度的绿地，其实也是个等待融进人们生活的秘密花园。有了这个临时设施，这个地方变成了公共空间，能为当地社区和游客所用。

犹太文化节日馆带有地域特征的屋顶景观隐藏在当地周围环境中，为人们提供了社交活动和自主活动的场所，将商业用途和非商业用途连为一体。BudCud 事务所设计了一个小型城市空间，不同建筑体量围绕着中央平台展开。中央平台可以用作舞台、舞池，也可以用作教室或者咖啡厅。三间房屋分别用作社区图书馆、Cheder 咖啡厅和来自特拉维夫的 Gelada 平面工作室。Gelada 工作室设有丝印工作坊。作为三间房屋的补充，还设计有三组台阶、长椅、坡道和一个乒乓球区。

所有结构都由木梁和厚胶合板搭建而成。三间房屋的外立面由天然胶合板和透明波纹塑料板装饰，与黄色地板明亮的色调非常协调。透明的墙体和屋顶不仅将阳光引入室内，也使每个体量的功能从外面一目了然。在晚上举办演唱会时，这些体量就成为引人注目的城市灯塔和富有表现力的布景素材。犹太文化节日馆项目在一个几乎被人们遗忘的公共广场上提供了一个建筑设施，从而使人们开始在这里进行社交活动。这一设计成为使城市生活变得更好的工具，挑战了空间使用的无限可能性。

FKZ Quarter

In 2015, anniversary 25th Jewish Culture Festival in Cracow celebrates Kazimierz – Jewish quarter of Cracow – as well as other Jewish districts from around the world. Following the theme, BudCud designed a temporary public space in one of a few empty squares in the district. Noticing the fact that the squares of Kazimierz are its most characteristic features, unfortunately all occupied by cars, monuments or commercial activities, with the festival architects decided to create the FKZ Quarter – an architectural device for bringing social life to the site. It is a temporary public urban situation created by multiple objects organised around a central elevated square. The FKZ Quarter is located on the back of the Old Synagogue,

.25

Kwartał
FKŻ
.25

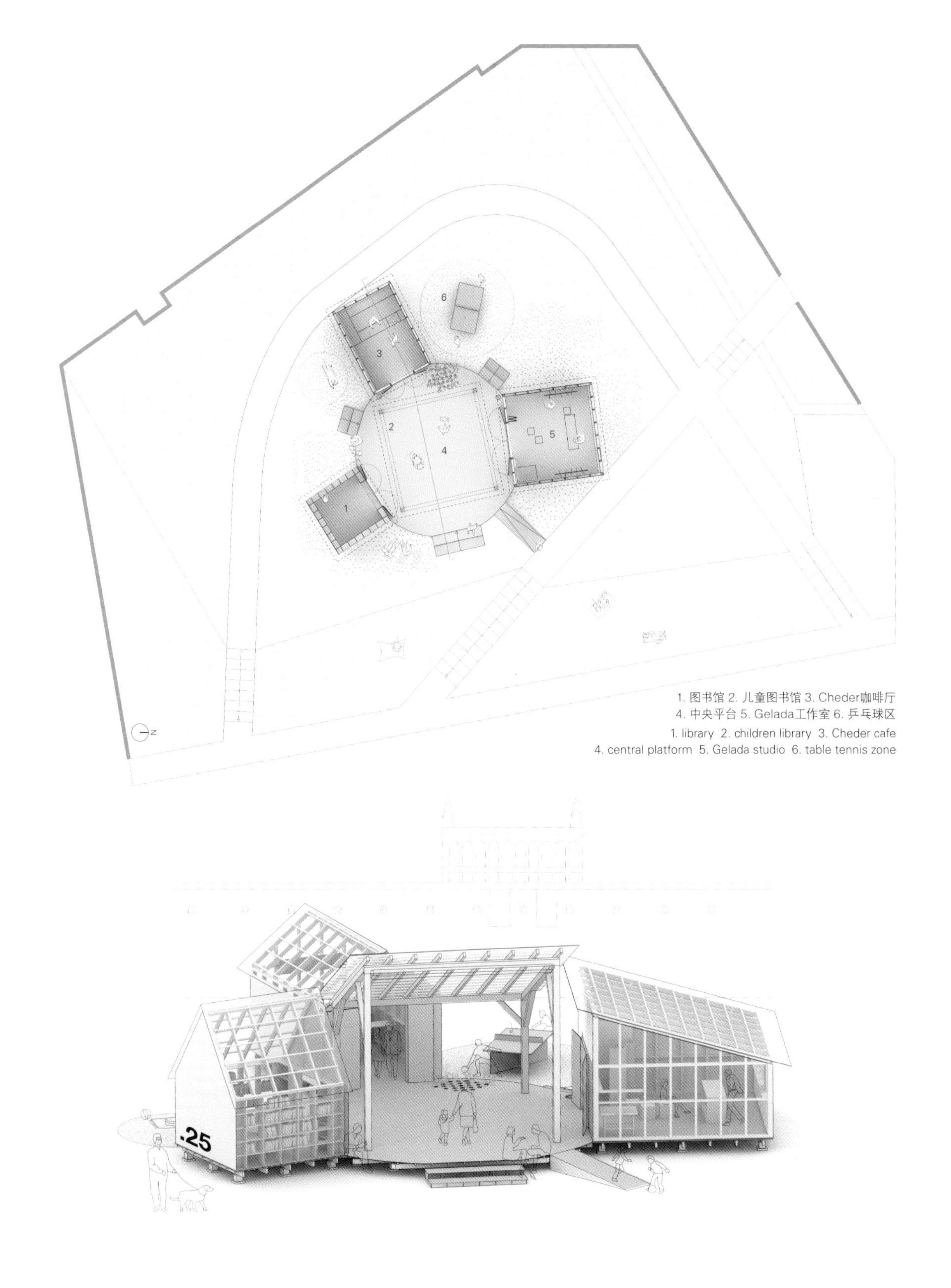

1. 图书馆 2. 儿童图书馆 3. Cheder咖啡厅
4. 中央平台 5. Gelada工作室 6. 乒乓球区
1. library 2. children library 3. Cheder cafe
4. central platform 5. Gelada studio 6. table tennis zone

at Dajwor Street, where no one goes, but bus drivers park tourist buses along the street. The site – green lawn with a natural slope – is a secret garden, waiting to be brought to life. With a temporary program the location was made public and used by local community and tourists.
The FKZ Quarter depicts architecture hidden in local circumstances with its vernacular roofscape, frames for social engagement and spontaneous behaviour, linking the commercial and the gratuitous. BudCud introduced a petite urban situation, where different volumes surround the central platform, that can become a stage, dance floor, classroom or a cafe. Three staircases – benches, a ramp and a table tennis zone complement three houses, operating as a social library, Cheder cafe and Gelada studio from Tel Aviv, where screen print workshops are held.
All the structure is made of wooden beams and thick plywood plates. The houses are finished with natural plywood and transparent corrugated plastic plates, balancing the bright colour of yellow floor. Transparent walls and roofs provide natural light inside, but also manifest the functions of singular volumes, which during night concerts become eye-catching urban lighthouses and expressive scenography.
The FKZ Quarter renders a possibility of social engagement by providing an architectural frame in forgotten public square. The design becomes a tool of living better in the city, challenging the possibilities of using the space.

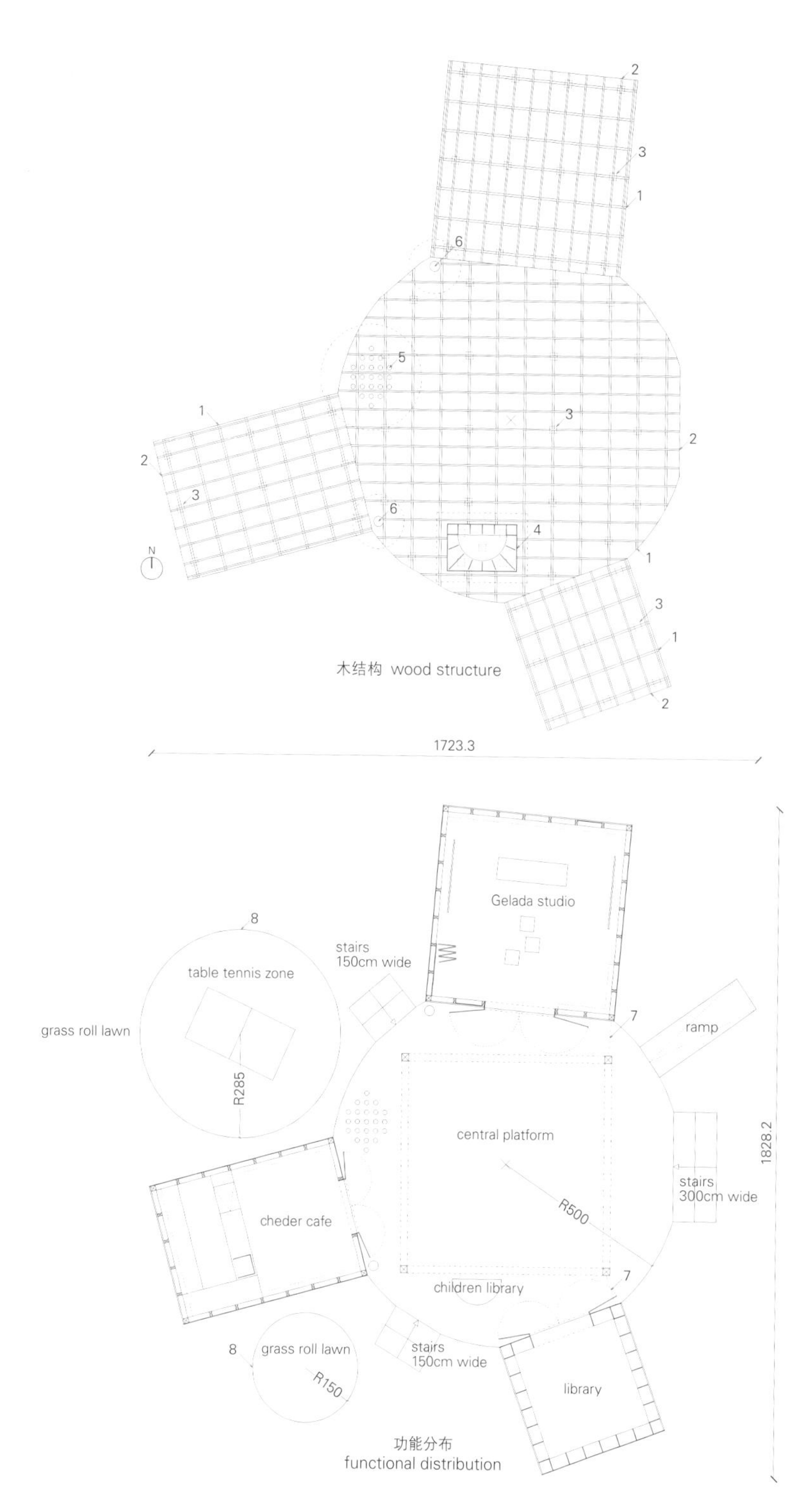

1. lower joist - wooden joist; section 6x16cm
2. upper joist - wooden joist; section 4x5cm
3. punctual support - wooden support; total height of 22.5cm, consisting of a wooden beam 20x20x17.5cm, a concrete flagstone 30x30x5cm, a horizontal insulation
4. children library
5. cut-outs for mint pots
6. cut-outs for trash bin
7. joint - a joint between the roof of central platform and pavilions of the library and Gelada studio; joint's type to be confirmed with the designer
8. grass - apply a grass roll on a previously prepared surface

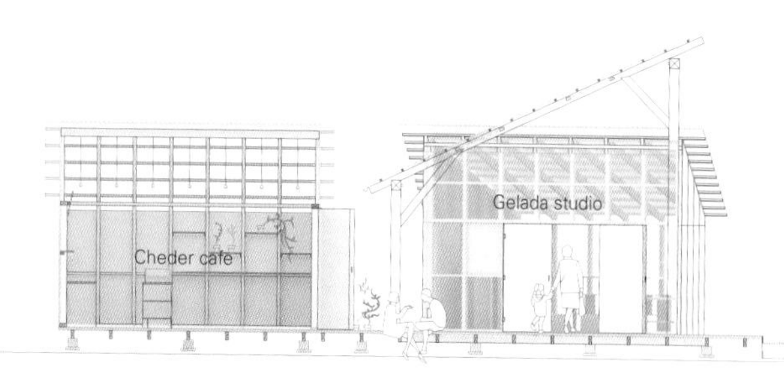

Czyteln
10:00 –
20:00
NON
STOP
P 24h

Gelada
GOLA & FKŻ CREW
GOLA
JOHN LEMON
JOHN LEMON

项目名称：FKZ Quarter / 地点：Cracow, Poland /建筑师：BudCud / 项目团队：Mateusz Adamczyk, Agata Wozniczka / 客户：Jewish Culture Festival in Cracow / 用途：pavilion, public space / 建筑面积：150m² / 结构：wooden frame / 材料：plywood, pvc transparent undulated sheets, wood
造价：EUR 20,000 / 竣工时间：2015 / 摄影师：©Haim Yafim Barbalat (courtesy of the architect)

当代艺术博物馆与城市规划展览馆

Coop Himmelb(l)au

CGN

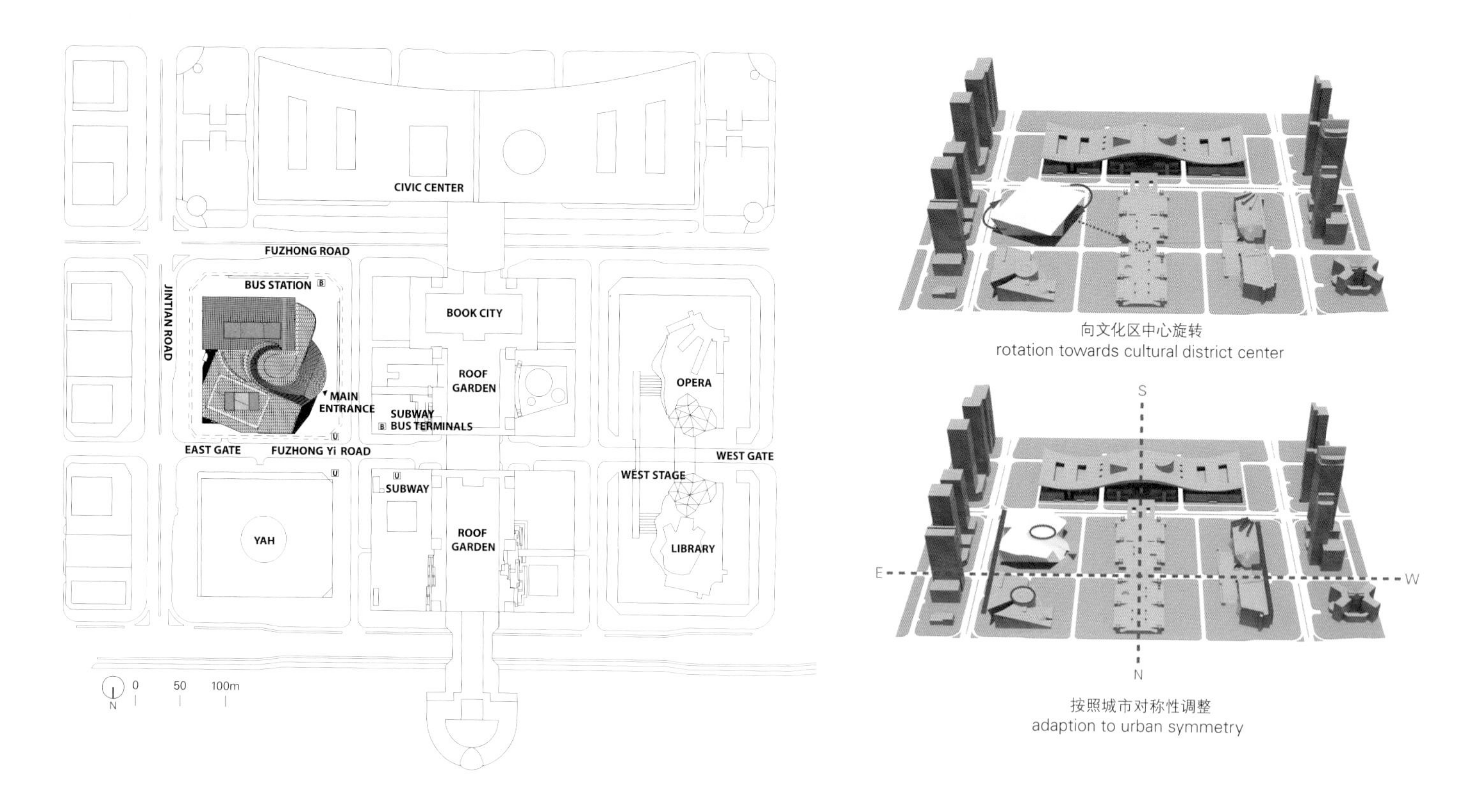

向文化区中心旋转
rotation towards cultural district center

按照城市对称性调整
adaption to urban symmetry

当代艺术博物馆与城市规划展览馆是深圳新城市中心福田文化区总体规划的一部分。这一项目包括两个处在同一建筑中却彼此独立的机构：当代艺术博物馆和城市规划展览馆。它们既是文化的汇聚地，又是建筑展览的场所。入口大厅、多功能展览大厅、礼堂、会议室和服务区都将为两个场馆所共用。两个场馆分别设计，强调各自在功能和艺术上的需求，二者又被统一在具有多功能立面的单一而庞大的建筑之下。透明的立面和复杂的室内照明系统使得两栋大楼间的公共入口和过渡区清晰可见。在室内，参观者对室外的城市景观一览无余，仿佛置身于有遮挡的室外，6～17m 高、完全开放的无柱展览场地更是加强了人们的这种体验。

在两个场馆之间的入口区的后面，参观者可乘自动扶梯或通过坡道向上到达主楼层，进入"广场"。这个广场也是参观博物馆的一个出发点，从这儿开始，可直接到达各个举办多种文化活动的房间、多功能厅、几个礼堂和一个图书馆。

一个银光闪闪、造型圆润的云形空间是整栋建筑的中心，也是"广场"的入口。"云"承载了多个楼层的许多公共功能，如咖啡厅、书店和博物馆商店，同时以连廊和坡道的方式连接了两个场馆的展厅。"云"以曲线形的表面，连通两个不同的空间，反映了两个场馆在同一屋檐下的构想。

当代艺术博物馆与城市规划展览馆的建成使得城市中心整体规划的东侧部分变得完整，填补了福田文化区北侧青年活动中心和南侧歌剧院－图书馆综合体之间的空隙。与这一地区其他建筑一样，当代艺术博物馆与城市规划展览馆主楼层高于地面 10m，由此营造出一个颇具舞台感的平台，将整栋建筑与周围建筑连为一起。

建筑的外表皮有两层：外层是由天然石材制成的百叶遮阳板，内层是具有微气候调节作用的隔热玻璃围护结构。以上两个元素构成了与建筑的框架结构分离的动感十足的表面。这一功能性外围护结构将两个场馆、一条垂直通道、娱乐区（"云"）、公共广场以及多功能地下空间统统包裹起来。

该建筑设计应用了高科技的建筑设备来降低建筑对外部能源的需求：无污染系统和设施都使用可再生能源，如太阳能、地热能（带有地下水冷却系统），全部安装高效节能系统。屋顶可以滤进阳光，为展厅提供了自然光，减少了人工照明的需求。当代艺术博物馆与城市规划展览馆集高科技的技术组件、紧凑的建筑体量、高效的隔热和遮阳系统于一身，不仅是一座建筑学意义上的地标，也是一个在生态和环保方面的标杆性项目。

Museum of Contemporary Art & Planning Exhibition

The Museum of Contemporary Art & Planning Exhibition (MOCAPE) is part of the master plan for the Futian Cultural District, the new urban center of Shenzhen. The project combines two independent yet structurally unified institutions: The Museum of Contemporary Art (MOCA) and the Planning Exhibition (PE) as a cultural meeting point and a venue for architectural exhibitions. The lobby, multifunctional exhibition halls, auditorium, conference rooms and service areas will be used jointly. Both museums are designed as separate entities emphasizing their individual functional and artistic requirements and yet are merged in a monolithic body surrounded by a multifunctional facade. This transparent facade and a sophisticated internal lighting concept allow a deep view into the joint entrance and transitional areas between the buildings. From the inside, visitors are granted an unhindered view onto the city suggesting they are somewhere in a gently shaded outdoor area, an impression enhanced by 6 to 17 meters high, completely open and column-free exhibition areas.

Behind the entrance area between the museums, visitors ascend to the main level by ramps and escalators and enter the "Plaza", which serves as a point of departure for tours of the museums. From the Plaza the rooms for cultural events, a

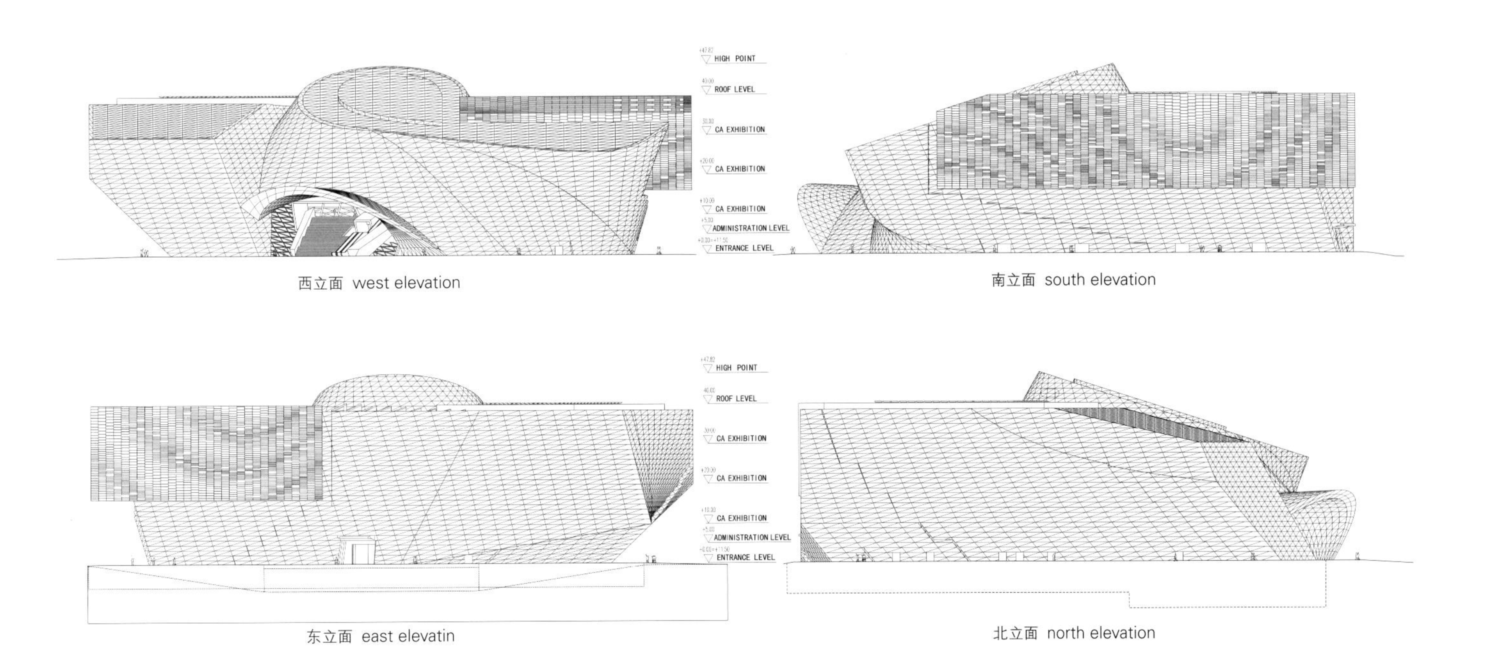

西立面 west elevation

南立面 south elevation

东立面 east elevatin

北立面 north elevation

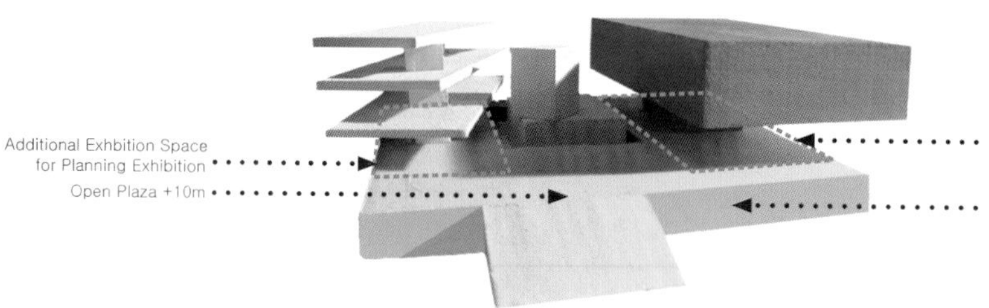

当代艺术展览馆盒式结构 Contemporary Art Exhibition box
城市规划展览馆平台 Planning Exhibition platforms
开放广场 open plaza

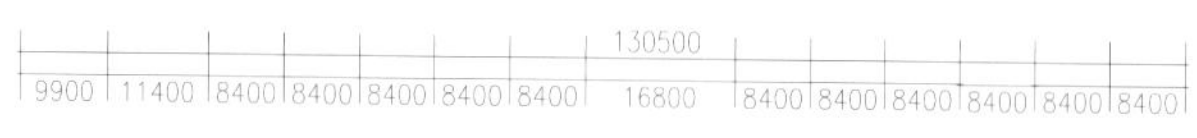

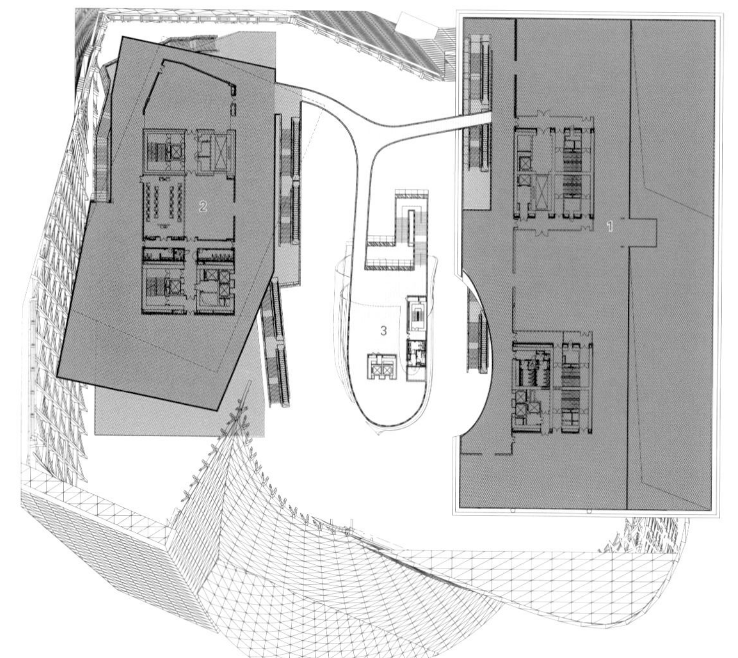

1. 当代艺术博物馆 2. 城市规划展览馆 3. 交流中心/书店/咖啡厅
1. Contemporary Art Exhibition Museum
2. Planning Exhibition Museum 3. communication center/bookshop/cafe
二层 second floor

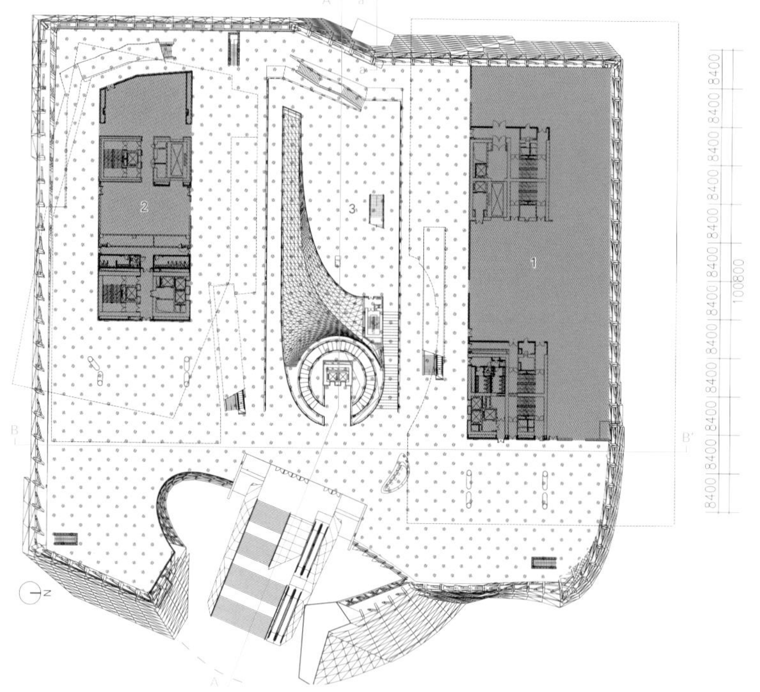

1. 当代艺术博物馆 2. 城市规划展览馆 3. 开放广场/多功能区域
1. contemporary art exhibition museum
2. planning exhibition museum 3. open plaza/multi-functional area
一层 first floor

项目名称：Museum of Contemporary Art & Planning Exhibition
地点：Shenzhen, China
建筑师：COOP HIMMELB(L)AU
规划师：COOP HIMMELB(L)AU – Wolf D. Prix & Partner ZT GmbH
首席设计师：Wolf D. Prix / 项目合伙人：Markus Prossnigg
设计建筑师：Quirin Krumbholz, Jörg Hugo, Mona Bayr
项目建筑师：Angus Schoenberger, Veronika Janovska, Tyler Bornstein
项目协调：Xinyu Wan
项目团队：Jessie Castro, Jessie Chen, Jasmin Dieterle, Luis Ferreira, Peter Grell, Paul Hoszowsky, Dimitar Ivanov, Ivana Jug, Zhu Yuang Kang, Alexander Karaivanov, Nam La-Chi, Rodelle Lee, Feng Lei, Megan Lepp, Samuel Liew, Thomas Margaretha, Jens Mehlan, Ivo de Nooijer, Reinhard Platzl, Vincenzo Possenti, Pete Rose, Ana Santos, Jutta Schädler, Günther Weber, Chen Yue
数字项目团队：Angus Schoenberger, Matt Kirkham, Jasmin Dieterle, Jonathan Asher, Jan Brosch
当地建筑师：HSArchitects
结构工程师：B+G Ingenieure, Bollinger und Grohmann GmbH
电气工程师：Reinhold Bacher
照明设计：AG Licht
客户：Davis Langdon & Seah
成本管理：Shenzhen Municipal Culture Bureau, Shenzhen Municipal Planning Bureau
用地面积：21,688m² / 有效楼层面积：80,000m²
建筑高/宽/长：40m/160m/140m
建筑规模：two stories below ground, five stories above ground
竞赛时间：2007 / 设计时间：2008 / 施工时间：2013—2016
摄影师：©Duccio Malagamba (courtesy of the architect)

出入口

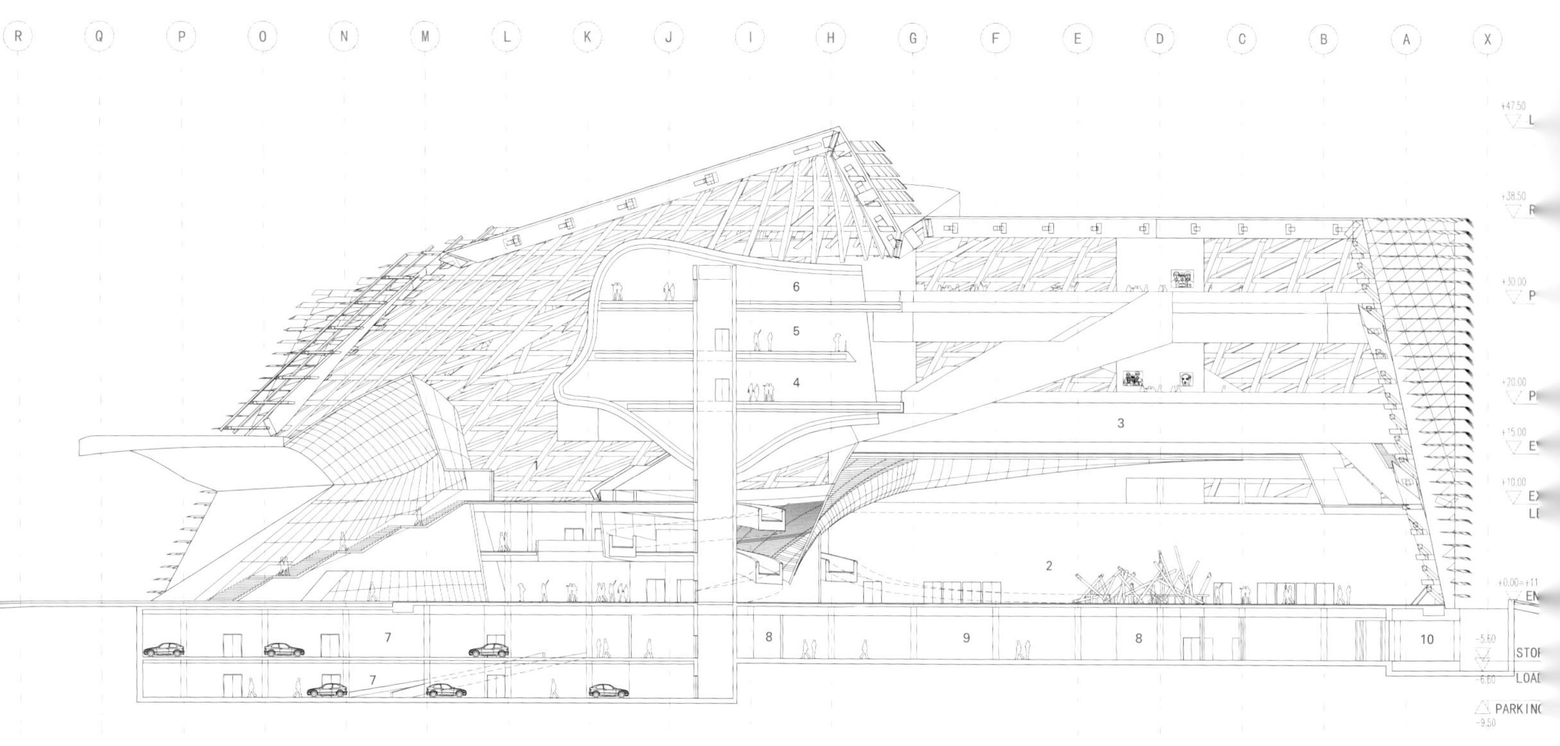

1. 大厅 2. 展览大厅 3. 上层大厅 4. 等候区/文化沙龙 5. 咖啡厅/艺术沙龙 6. 茶室/城市规划展览馆艺术沙龙
7. 地下停车场 8. 技术室 9. 临时储藏室 10. 装卸区
1. lobby 2. exhibition hall 3. upper lobby 4. waiting area/culture salon 5. cafe/art salon 6. tea house/PE art salon
7. underground garage 8. technical room 9. temporary storage 10. uploading and downloading area
A–A' 剖面图 section A-A'

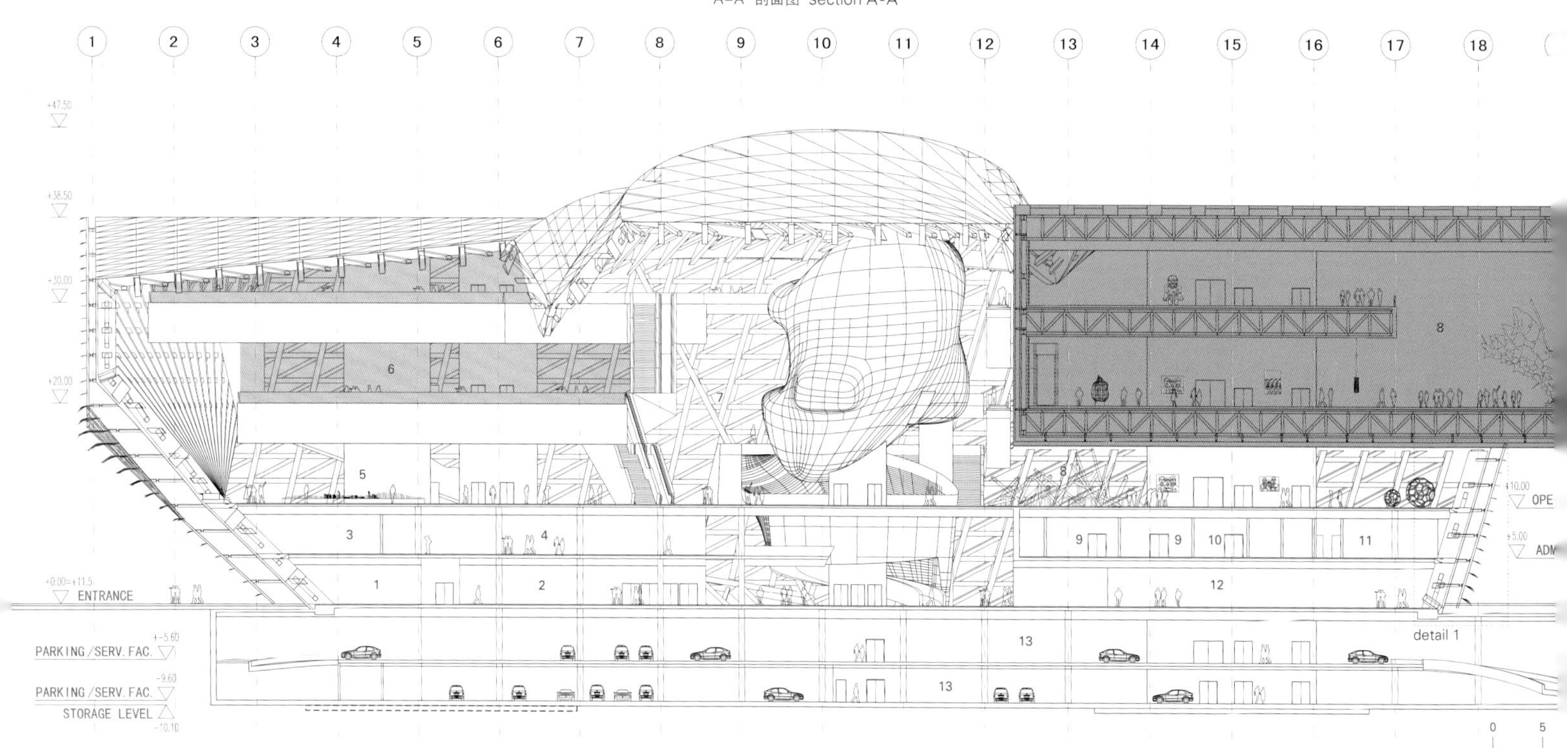

1. 小商店 2. 艺术商店 3. 行政管理部门接待处 4. 图书馆查询处/阅览室 5. 休息等候区 6. 城市规划档案开放展区
7. 大厅 8. 公众大厅 9. 会议室 10. 电话室 11. 行政管理办公室 12. 餐厅/咖啡厅 13. 地下停车场
1. small shop 2. art shop 3. administration reception 4. library information/reading room 5. resting and waiting area 6. open exhibition of planning achivements
7. lobby 8. general hall 9. meeting room 10. telephone room 11. administration 12. restaurant/cafe 13. underground garage
B–B' 剖面图 section B-B'

multi-functional hall, several auditoriums and a library can be accessed.

A silvery shining and softly deformed "Cloud" serves as a central orientation and access element on the Plaza. On several floors the Cloud hosts a number of public functions such as a café, a book store and a museum store and it joins the exhibitions rooms of both museums with bridges and ramps. With its curved surface the Cloud opens into the space reflecting the idea of two museums under one roof.

The MOCAPE monolith completes the eastern part of the master plan for the city center and fills the last gap in the Futian Cultural District between the "Youth Activity Hall" (YAH) to the north and the opera-library complex to the south. Similar to other buildings in this district, the main level of the MOCAPE lies 10 meters above the ground level and so creates a stage-like platform, which acts as a unifying element with the adjacent buildings.

The exterior skin consists of an outer layer of natural stone louvers and the actual climate envelope made from insulated glass. These elements form a dynamic surface, which is structurally independent from the mounting framework of the museum buildings. This functional exterior envelops the two museums, a vertical access and entertainment element (Cloud), the public Plaza, and the multifunctional base.

The technical building equipment is designed to reduce the overall need of external energy sources: Pollution free systems and facilities use renewable energy sources through solar and geothermal energy (with a ground water cooling system) and only systems with high energy efficiency have been implemented. The roof of the museum filters daylight for the exhibition rooms, which reduces the need for artificial lighting. With this combination of state of the art technological components, a compact building volume, thermal insulation and efficient sun shading the MOCAPE is not only an architectural landmark but also an ecological and environmentally friendly benchmark project.

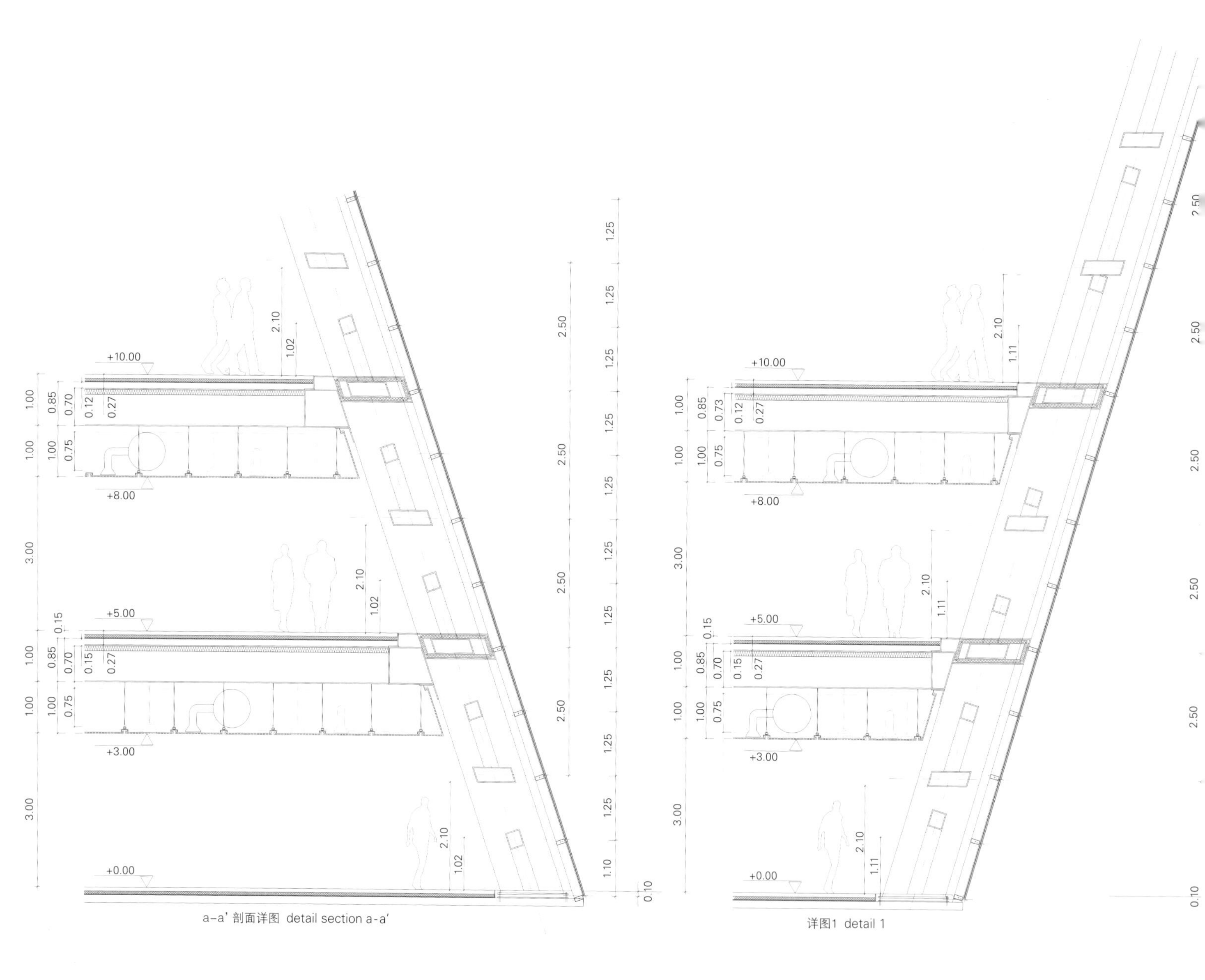

a–a' 剖面详图 detail section a-a'

详图1 detail 1

阿纳姆中央火车站

UNStudio

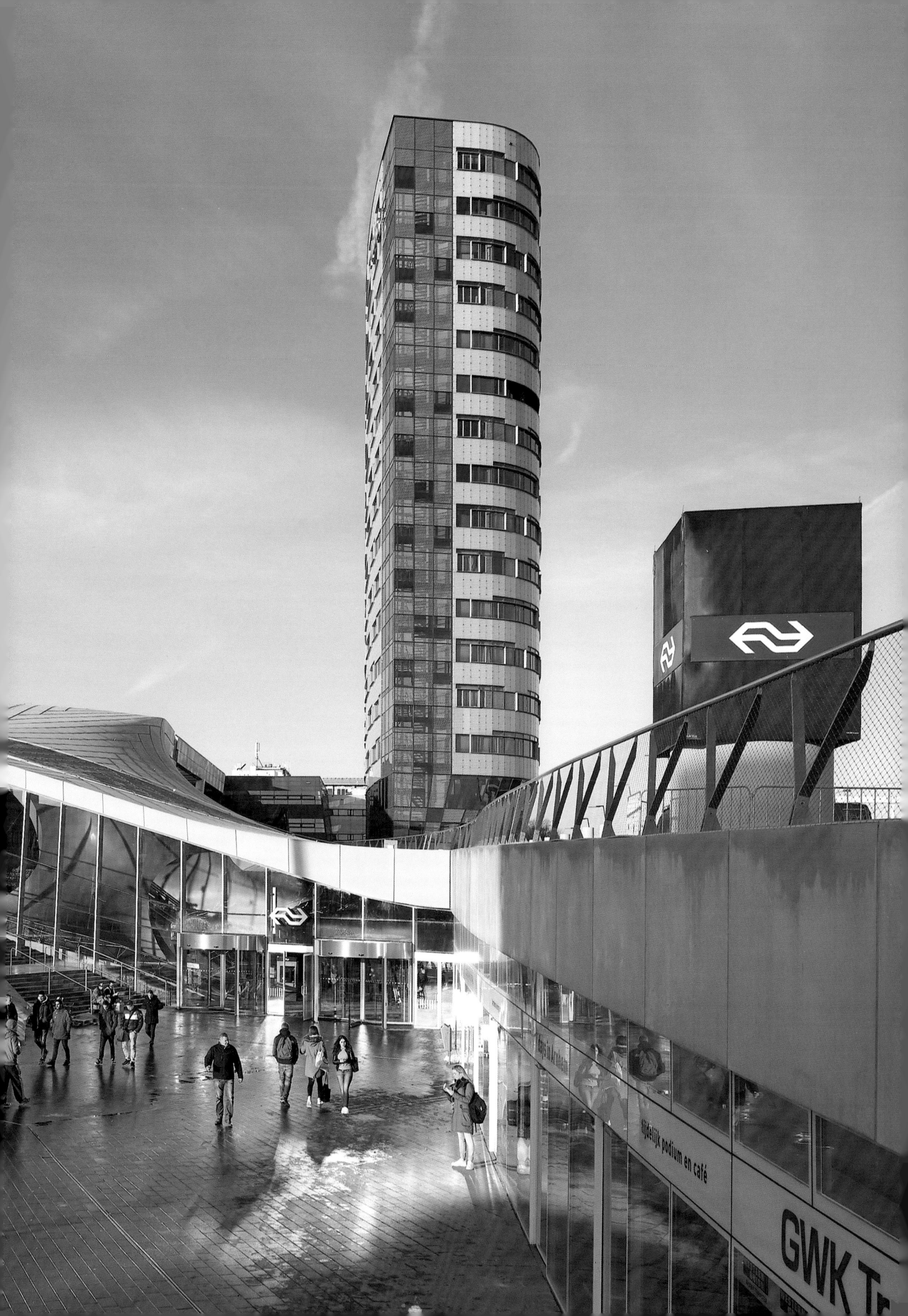
tijdelijk podium en café
GWK Tr

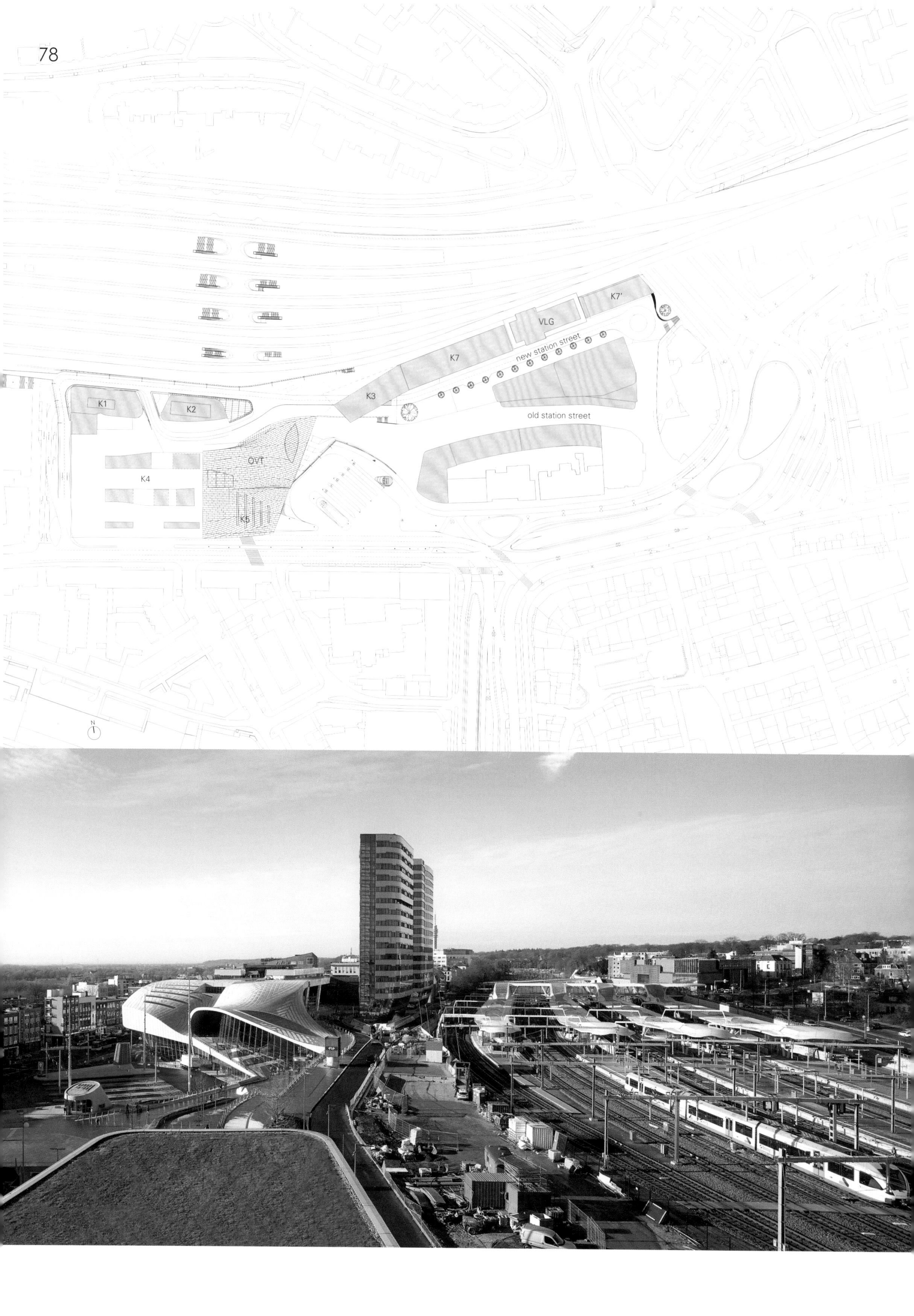
K7'
VLG
new station street
K7
K3
K1
K2
old station street
OVT
K4
K5
N

阿纳姆火车站从最初的设计图纸到建立完成共用了19年的时间。荷兰铁路局、设施管理局、基建环保部、市议会、UNStudio 建筑事务所以及承建方 Ballast Nedam-BAM 深度合作，共同建造出一个顶级标志性公共交通枢纽，使荷兰成为世界的焦点。

20 世纪末，阿纳姆市议会想要重建火车站地区。与此同时，依据荷兰国家重点项目体系，基建环保部也表达了修整包括阿纳姆在内的重要车站枢纽的野心，以满足旅客人数增加的需要，同时加强国内外铁路交通运输。这一项目为作为项目组织机构的铁路局带来巨大挑战。

建筑师本·范·伯克尔设计的这个公共交通枢纽充满了未来感：一个绝妙的能刺激人一切感官的屋顶，流畅的曲线形状，精美的木结构，还有宽阔的旅客通道。这是一栋每天可接待 11 万人次的宏伟建筑，拥有浑然天成的线条和将所有结构汇聚于一体的一个独一无二的元素，有着扭曲缠绕的前脸。这是一栋看起来扭曲缠绕的建筑，空间得到拓宽的同时也成为一个聚集点，引导旅客去往各自的目的地。这是一个让人感到惊艳的火车站，初来此地感觉如此，再次往返亦是如此。由于建筑工程技术十分复杂，这个项目分两阶段进行。2011 年，新自行车存放处和车站平台隧道完成。同时，铁路局的另一个项目（阿纳姆市轨道项目）改进了阿纳姆附近的铁路基础设施，包括第四个站台及新站台屋顶的施工。2012 年，承建方 BAM 不惧挑战，负责阿纳姆火车站项目的重点任务，即公共交通枢纽的建设。

通过先进的 3D 技术以及铁路负责人、承建方和分包商三方的创新合作，车站的上部构件采用重量轻、安装快、质量优的造船用钢代替了混凝土。项目取得了成功，于 2015 年 11 月 9 日为阿纳姆增添了一座独一无二的地标性建筑。

新火车站连通 Sonsbeek 公园和阿纳姆市的市中心，是进入荷兰的重要门户。建筑师本·范·伯克尔在这里实现了他的创新理念，但是这不仅仅是个设计，它还向每一位旅客和游客承诺，能够随时指引人们到达他们想要去的地方。自然的高度差、庞大透明的外立面和天窗的使用都使这个公共交通枢纽易于被人们接受和理解。例如，火车站、公共汽车站、停车场、自行车存放处、入口和出口等主要目的地都非常容易识别。经过多年的建设，阿纳姆和它的居民们终于能够享受这个已经世界著名的超现代车站所带来的种种益处了。

Arnhem Central Transfer Terminal

Arnhem Station has been completed 19 years after the initial drawings were made. The deep collaboration between ProRail, NS, the Ministry of Infrastructure & the Environment, the Municipality of Arnhem, UNStudio and the construction consortium Ballast Nedam-BAM has culminated in an iconic public transport hub putting the Netherlands in the international spotlight.

At the close of the preceding century, the Municipality of Arnhem wanted to renew the station area. At the same time the Ministry of Infrastructure & the Environment expressed the ambition to, in the framework of the Nationale Sleutelprojecten (national key projects) modify important station hubs – including Arnhem – to keep pace with the number of travellers and, in turn, reinforcing national and international train transport. This presented a huge challenge to ProRail as the project organisation.

Architect Ben van Berkel came up with a futuristic plan: a public transport terminal with a sensational roof, fluid shapes,

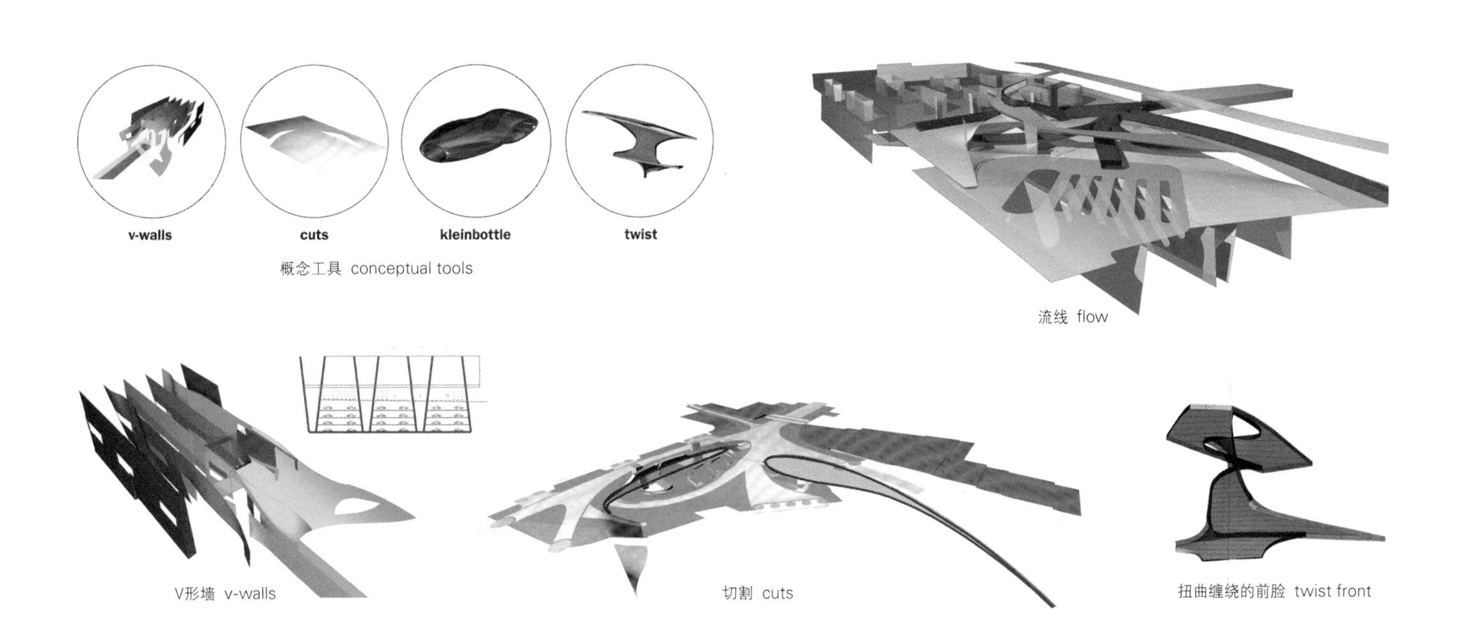

概念工具 conceptual tools

流线 flow

V形墙 v-walls

切割 cuts

扭曲缠绕的前脸 twist front

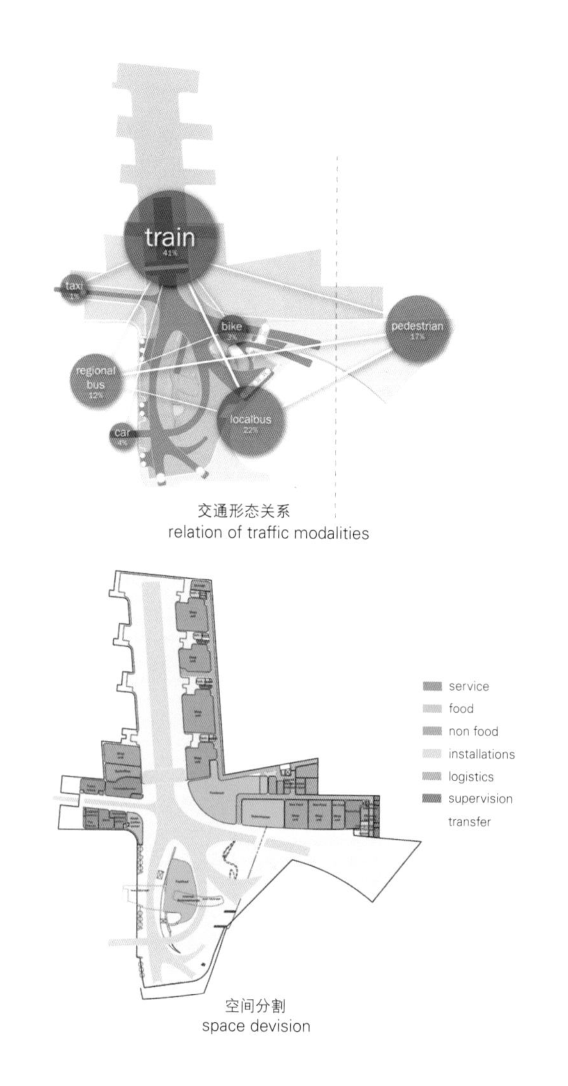

交通形态关系
relation of traffic modalities

空间分割
space devision

一层 ground floor

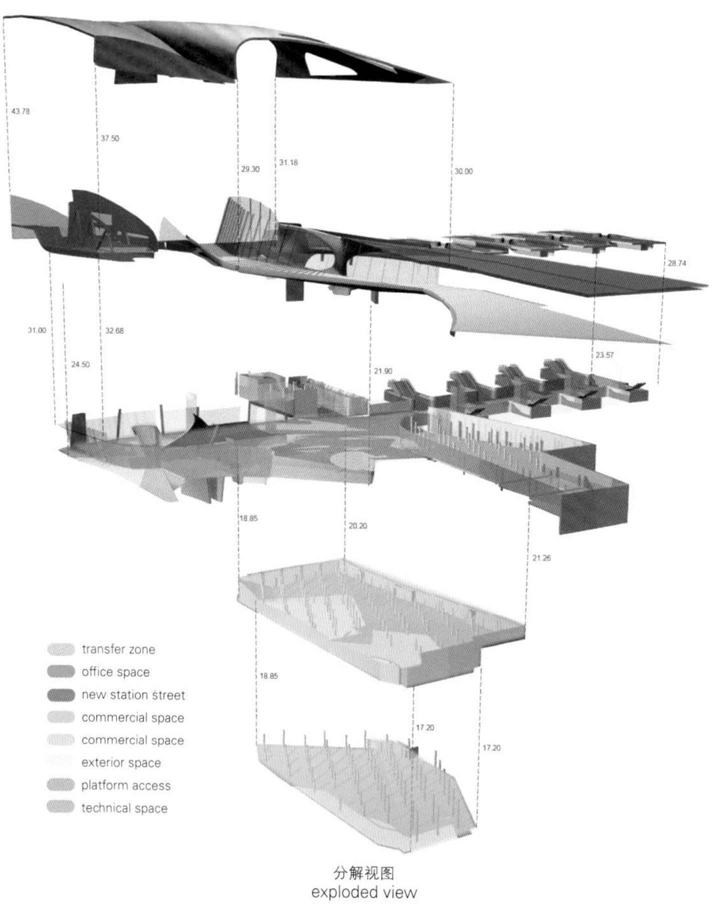

分解视图
exploded view

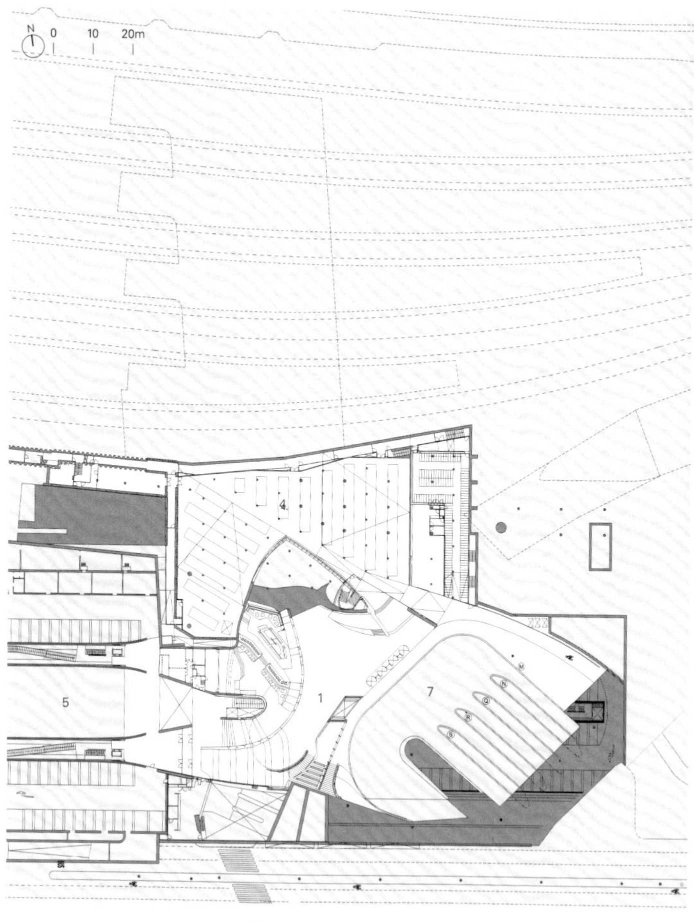

地下一层 first floor below ground

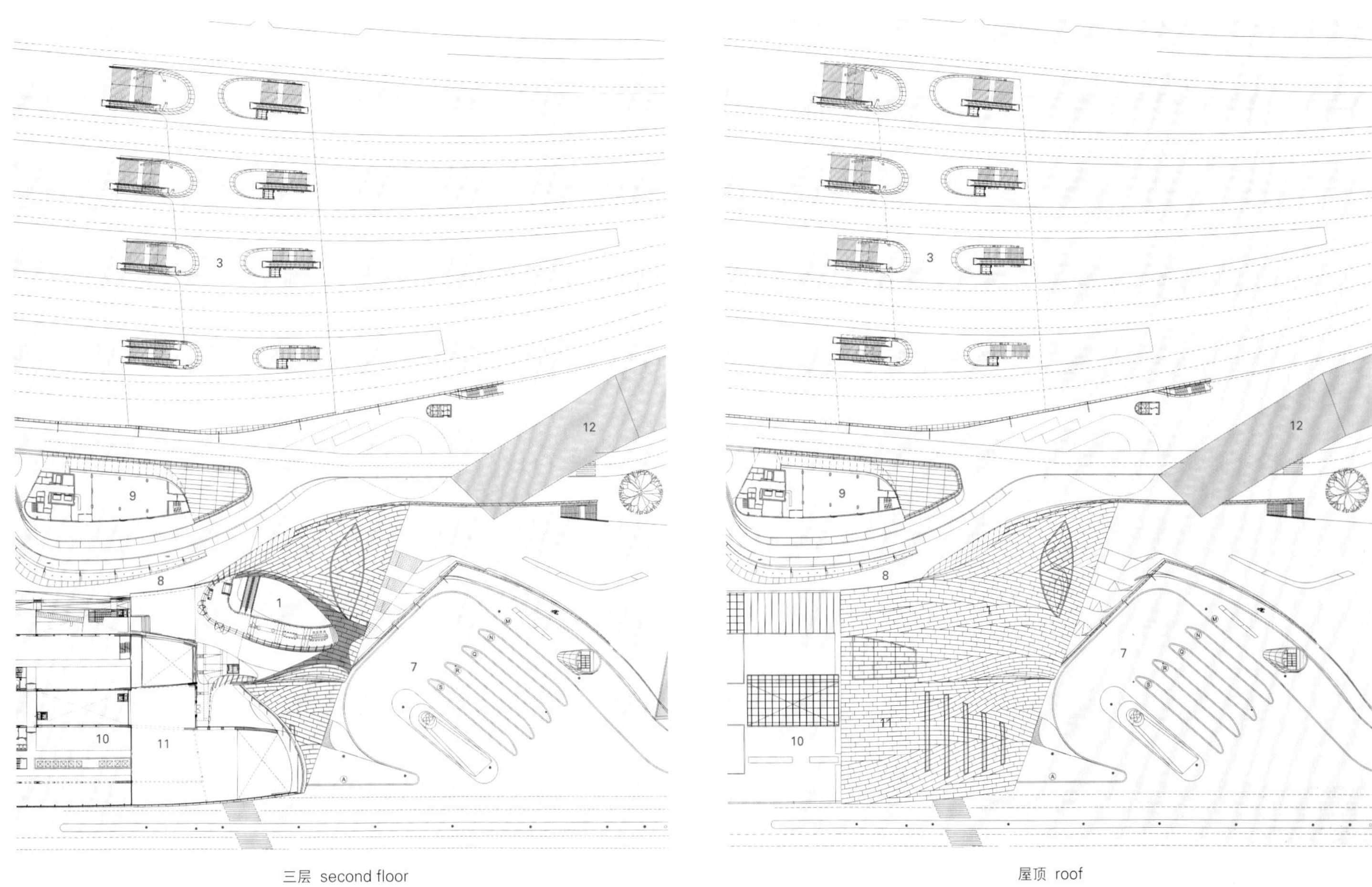

三层 second floor

屋顶 roof

1. 公共交通站点 2. 站台隧道 3. 站台&站台屋顶 4. 自行车存放处 5. 地下停车场 6. 公共汽车站（区域公交） 7. 公交广场（当地公交）
8. 抬高的办公广场 9. 办公塔楼（K2） 10.水平办公楼（K4） 11. 水平办公楼（K5） 12. 办公塔楼（K3，未建设）

1. public transport terminal 2. platform tunnel 3. platforms & platform roofs 4. bicycle storage 5. underground parking garage 6. bus terminal (regional busses) 7. bus square (local busses)
8. elevated office square 9. office tower (K2) 10. horizontal offices (K4) 11. horizontal offices (K5) 12. office tower (K3, future development)

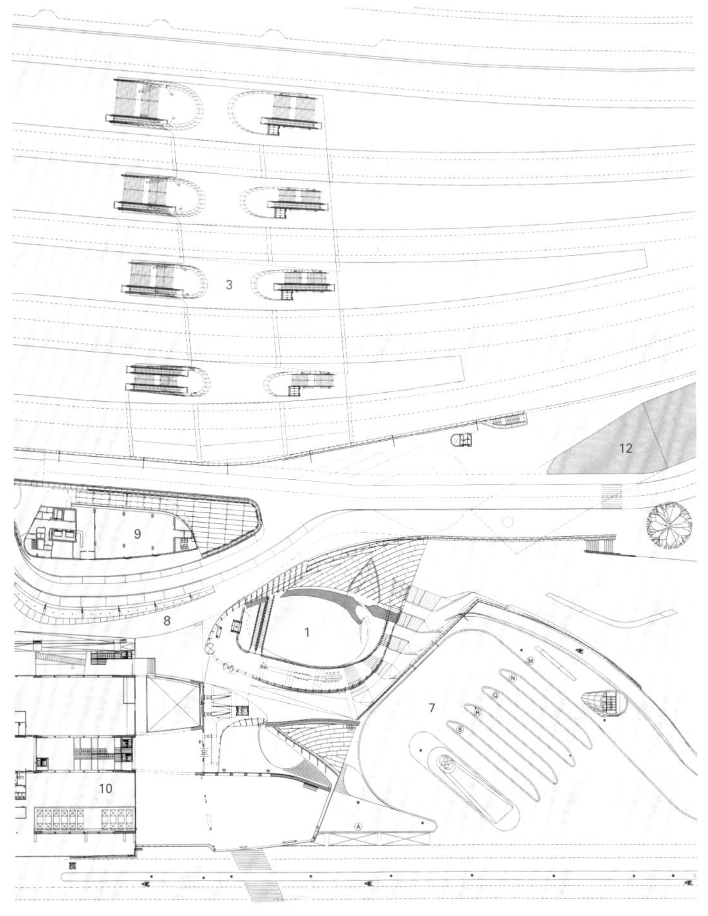

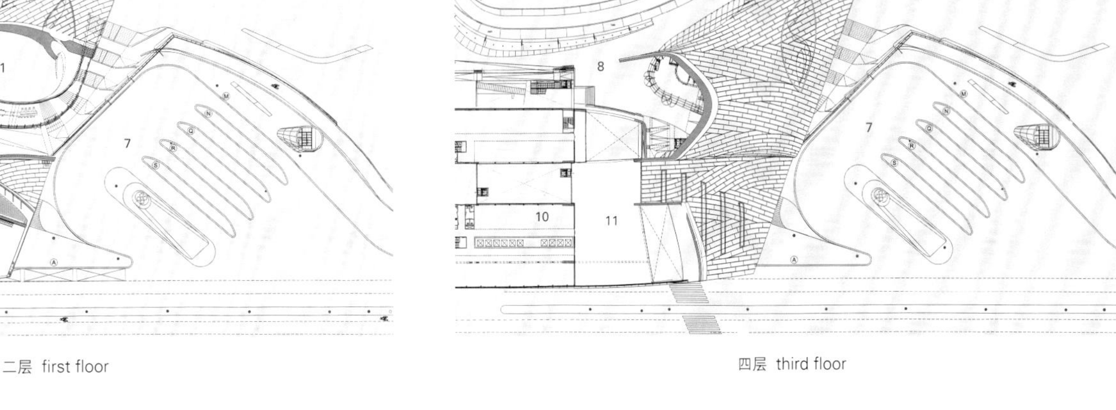

二层 first floor

四层 third floor

Bushaltes
M t/m S
P+R
Centrum

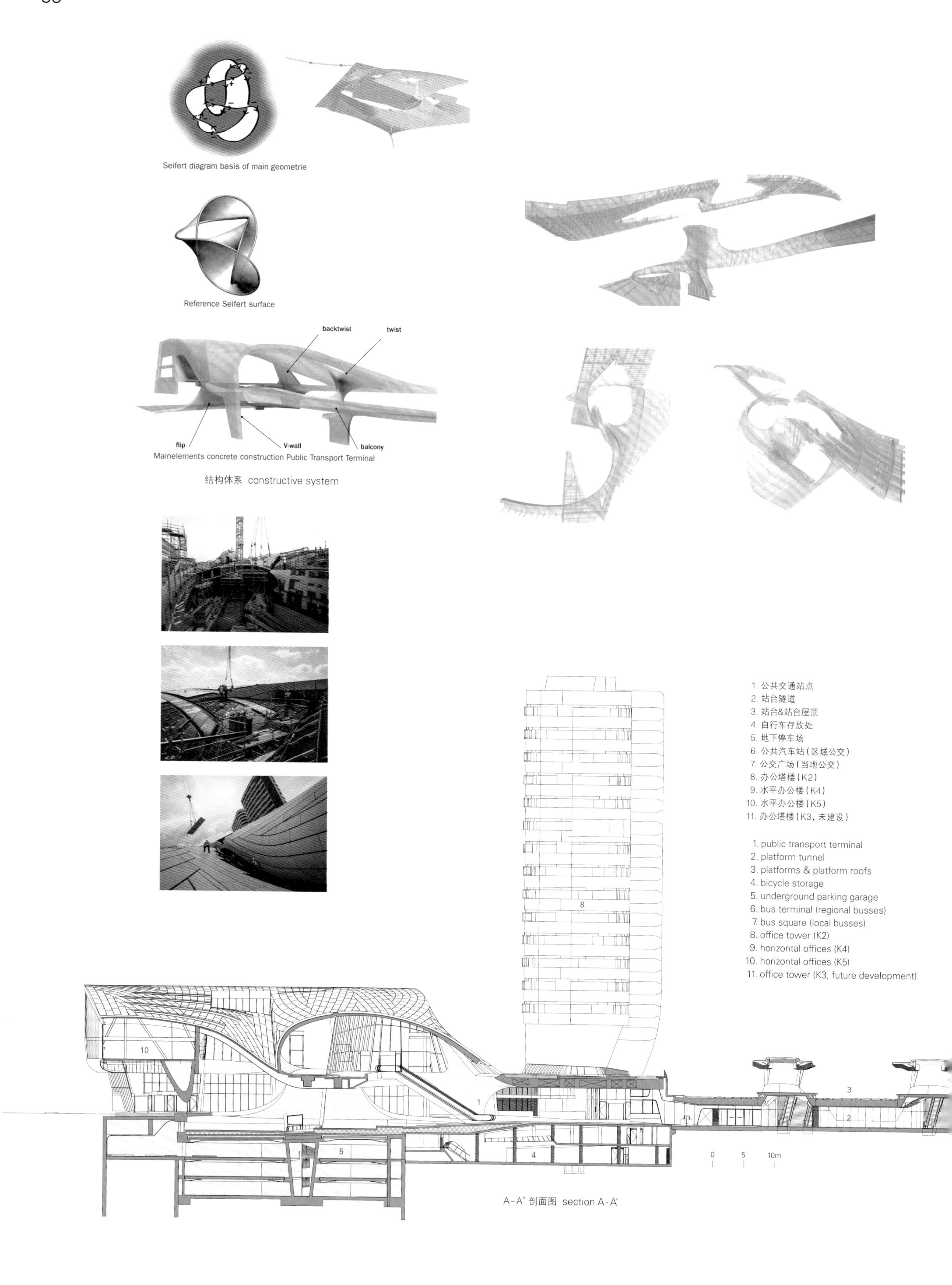

结构体系 constructive system

A-A' 剖面图 section A-A'

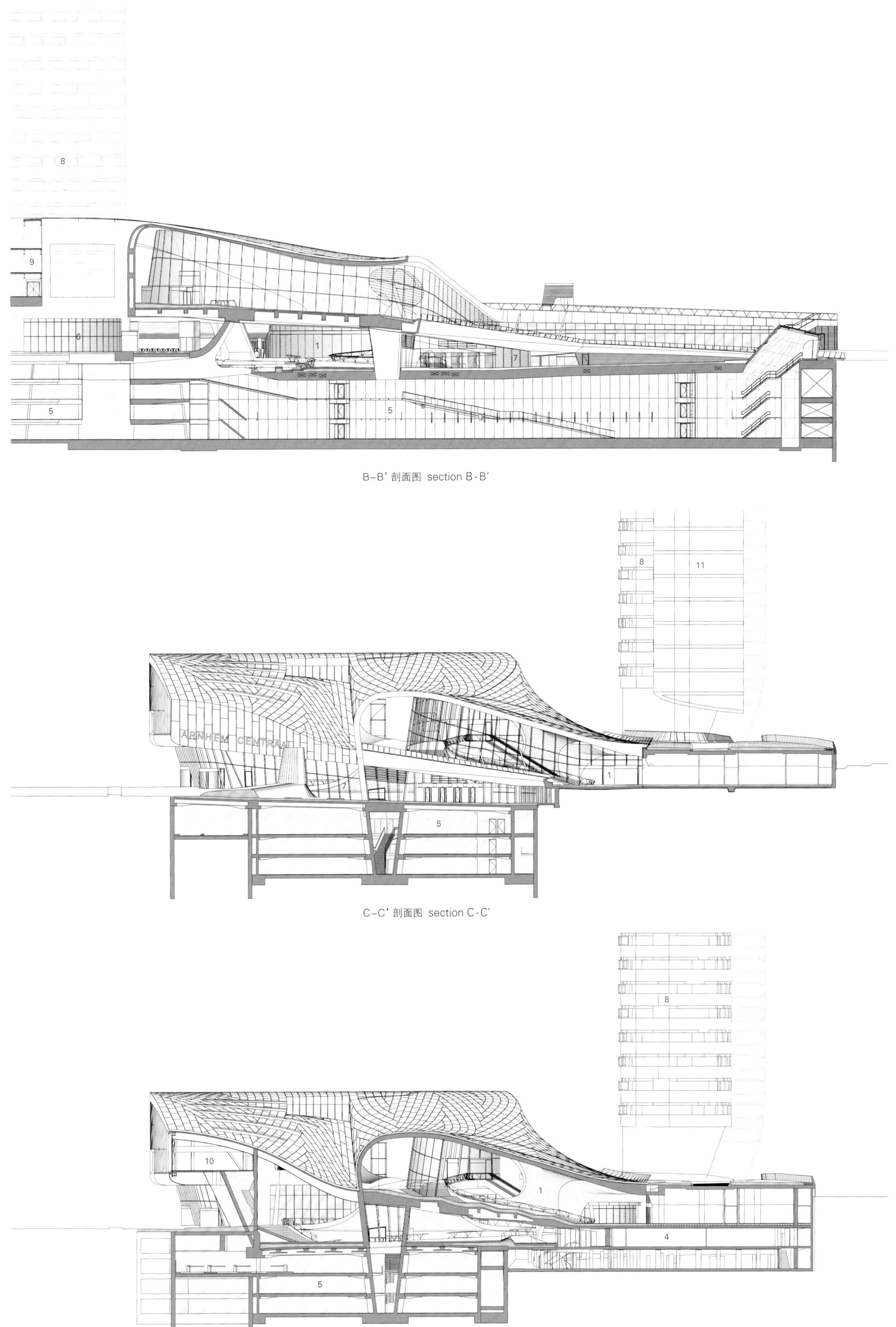

B–B' 剖面图 section B-B'

C–C' 剖面图 section C-C'

D–D' 剖面图 section D-D'

beautiful woodwork and a wide passenger tunnel. An edifice that can handle 110,000 travellers a day, with organic lines and a unique element where everything comes together: the front twist. A twisted shape that opens up the space and simultaneously constituting a meeting point and helping people find their way. A station that makes you catch your breath, whether you are there for the first time or returning. Because construction was going to be technically complex it was decided to tackle the project in two phases. In 2011, the new bicycle storage and platform tunnel were completed. At the same time, another ProRail project – Sporen in Arnhem (tracks in Arnhem) modified the rail infrastructure around Arnhem. This included the construction of a 4th platform and new platform roofs. In 2012, the construction consortium Ballast Nedam – BAM had the guts to take on the keystone of the Arnhem station project, the construction of the public transport terminal.

Instead of concrete, the terminal's upper elements were built of shipbuilding steel which is lighter, faster to erect and of higher quality. This demanded advanced 3D techniques as well as innovative collaboration between the principal, the construction consortium and sub-contractors. The project was a success and on 19 November 2015, Arnhem gained a unique landmark.

The new station links Park Sonsbeek and Arnhem's city center and acts as an important gateway to our country. Van Berkel's exceptional ideas came to life here, but it isn't just about design. Being able to see where you need to go at a glance is the promise given to every traveller and visitor. Natural differences in elevation and the use of sizeable transparent facades and skylights make the public transport terminal easy to comprehend. For instance, the main destinations: the railway and bus stations, the car park, the bicycle storage and the entrances and exits are all very recognisable. After years of construction, Arnhem and its inhabitants can now reap the benefits of a hyper-modern station that is already world famous.

项目名称：Arnhem Central Transfer Terminal / 地点：Stationsplein 38, Arnhem, The Netherlands / 建筑师：UNStudio / 施工管理：ProRail B.V. with back-office support of multiple parties / 结构工程师：Arup Amsterdam (public transport terminal), Van der Werf & Lankhorst (bus station, parking garage and office square)_Design and tender phase; Arcadis (tender design)_Structure of pedestrian tunnel; BAM Advies & Engineering, ABT_Public transport terminal phase 2 / 机电管道设施：Arcadis_Design and tender phase; BAM Techniek, Unica_Public transport terminal phase 1 & Finishes of pedestrian tunnel; BAM Techniek, Unica_Public transport terminal phase 2 / 防火与生命安全：DGMR Bouw BV / 照明：Arup Lighting / 路径设计：Bureau Mijksenaar / 标书制定：ABT / 船舶建造：Centraal Staal, Groningen_Public transport terminal phase 2 / 总承包商：Besix-Welling_Structure of pedestrian tunnel; Bouwcombinatie BAM Ballast Arnhem, Centrum VOF (BBB, BAM & Ballast Nedam)_Public transport terminal phase 1 & Finishes of pedestrian tunnel; Bouwcombinatie BAM Ballast Arnhem, Centrum VOF (BBB, BAM & Ballast Nedam)_Public transport terminal phase 2
客户：ProRail B.V. / 用途：public transport terminal / 用地面积：11,250m² / 有效楼层面积：21,750m² / 建筑体积：90,000m³ / 建筑高度：20m above main street level, 27m above lowest floor level / 建筑规模：two stories below ground, four stories above ground / 结构：architectural exposed concrete_phase 1; architectural exposed concrete, coated (monocoque) steel structure (prefabricated modules)_phase 2 / 室外饰面材料：Aluminum & galvanized steel (custom systems)_substructure; Architectural exposed concrete, Coated monocoque steel structure & steel frameworks_superstructure / 设计时间：1998—2006 / 施工时间：2007—2015.10 / 开放时间：2015.11
摄影师：©Eva Bloem (courtesy of the architect) - p.90; ©Hufton+Crow (courtesy of the architect) - p.76~77, p.78, p.79, p.80, p.84~85, p.86, p.92~93, p.94, p.95; ©Ronald Tilleman (courtesy of the architect) - p.87, p.91

公共建筑；城市社会基础设施

Public B
Urban S
Infrast

本质上，城市是人们聚集的地方。然而，在过去的一个世纪中，新技术的发展和朝私有化发展总的趋势将城市的聚集功能从必需变成了一个非必需的选择。以前公共空间中进行的活动，现在越来越多地在私人空间或半公共空间中进行。由于公共空间的用途具有了可选择性，其质量和吸引力也相应地变得更加重要。

In its essence, the city has always been a meeting place. In the past century, however, new technologies and a general trend towards privatization have contributed to changing this meeting function from a necessity to a choice. Activities that previously took place in public spaces now increasingly take place in (semi-) private spaces. As the use of public spaces becomes optional, their quality and attraction become correspondingly more important.

ildings; ocial ructure

公共建筑是如何吸引人们在此消遣，从而提升其作为城市人们聚集之所的功能的呢？纵观最近的一些项目，可以看到公共建筑是如何通过提供一系列娱乐活动、通过创建欢迎人们使用的内部公共空间、通过与其周边环境的相融相通而提供各种活动项目的。公共建筑正是以这样的方式成为城市社交基础设施，成为人们聚集之所。

How can public buildings contribute to the city's function as a meeting place, making it into a place where people choose to spend their time? An overview of recent projects demonstrates how public buildings are generating activity by offering a mix of recreational programs, by creating public interiors that invite use, and by making connections with their context. In these ways, public buildings act as anchors of the urban social infrastructure.

雅高酒店竞技场馆，法国
Accorhotels Arena, France

说到公共建筑，人们脑海里浮现的是装潢材料考究、超大空间比例的宏伟大厅以及城市生活的生动与活跃——人们在此办公、会友，或者只是在此看看周围发生了什么。事实上，城市一直扮演着交流场所的重要作用。城市成为人们会面和社交的场所，成为商品和服务的交易市场，成为各种城市功能间的纽带和桥梁。[1] 这种交流衍生出了一种社会凝聚力，随之而来的是一种归属感，这是参与社区生活意愿的基础，也是创建公益事业的基础。

虽然这种社交历来是城市日常生活中的自然现象，但如今人们越来越少地自发选择与他人在公共场合会面。现如今，随着私人住宅、汽车、办公室、电脑和通信技术的发展，曾经在公共领域进行的活动都可以在私下完成。[2] 购物、社交、分享想法和体验都可以在网上完成。

而且，随着从工业社会步入今天的消费型和休闲型社会，曾经促使人们进入公共领域的那些必要的、与工作相关的事情现在占用着人们越来越少的时间。随着生活水平不断提高，年轻人工作时间更长、成家更晚；不断完善的医疗保健服务使人们的寿命增加、退休后的时间也变多了。总体而言，人们有更多的时间和资源用于消费和消遣。[3]

随着私有化的不断发展以及财富和空闲时间的增加，公共空间更是一个可选可不选的地方，所以就出现了"选择型城市生活"。如今，参与公共生活不再是自发的，而是成为一种选择。当公共空间的质量不足以吸引人们的时候，街道和广场开始被人们遗弃，变得不安全，从而出现了"被遗弃的城市"这一现象。[4]

The experience of public buildings evokes images of grand halls with refined materials and noble proportions, and the liveliness and activity of city life – people going about their errands, meeting others, or simply watching what is happening around them. The essential role of cities, in fact, has always been as places of communication. They grew as places for meeting and social exchange, as marketplaces for the exchange of goods and services, and as connections between functions.[1] It is through this communication that a sense of social cohesion is born. With it arises a sense of belonging, which is the foundation for the will to participate in the life of communities, and to contribute to building a public good.

While this social exchange has historically been a natural consequence of daily urban life, however, meeting others in public is gradually becoming less automatic. With today's private dwellings, cars, offices, computers and communication technology, the activities that once took place in the public realm can now be done in private.[2] Shopping, socializing, and sharing one's opinion and experiences can all be done online.

As well, with the shift from an industrial society to today's consumer and leisure society, the essential, work-related errands that once drew people into the public realm take up much less of our time. Living standards have improved, young people work longer and start families later, and improved healthcare has increased longevity and the years after retirement. Overall, people have more time and resources for consumption and pleasure.[3]

Because of increasing privatization, wealth and free time, the use of public space has largely become optional, giving rise to an "elective city life". Participation in public life is no longer automatic; today, it is a choice. When public spaces are not of sufficient quality to attract use, streets and squares become deserted and unsafe, resulting in "abandoned cities".[4]

犹太文化节日馆，波兰
FKZ Quarter, Poland

阿纳姆中央火车站，荷兰
Arnhem Central Transfer Terminal, The Netherlands

如果城市继续将公共建筑的主要功能定位为社交场所的话，它们该如何适应社会的发展？公共建筑如何能给人们一个来参观这个城市的理由？如何吸引人们在此会面？在公共建筑中进行的活动该如何延展至周围的公共空间，从而打造一个富有活力的公共区域？通过概览本书所选择的最近几年的项目设计，我们可以看到公共建筑设计师们是如何应对上述问题的。

1. 吸引活动：设置公共项目

全新的"选择型城市生活"给城市带来了全新的、扩展的城市功能设施，其中包括用来举办文化活动、展览和体育赛事的设施。大型的用来举办不同活动的建筑项目有占地面积达 62 000m² 的巴黎雅高酒店竞技场馆（由 DVVD 事务所设计）。巴黎雅高酒店竞技场馆可以容纳 7000 到 20 000 名观众，可以举办音乐会和各种不同的国际体育赛事。

文化类建筑项目包含的活动更加多种多样，因此要求其建筑空间更加灵活多变。比如，扩建后的斯特拉斯堡会议中心项目（由 Dietrich | Untertrifaller 与 Rey-Lucquet et associés 联合设计）包含六个大小不一的礼堂和会议厅、15 间会议室，还有多间酒吧、一个餐厅以及办公空间。加上邻近的展览园，这个会议中心可以举办大型会议、公司企业会议、展览和文化活动。

由 BudCud 事务所设计的犹太文化节日馆位于克拉克夫市犹太人社区卡齐米日。设计师通过将设计元素围绕一个中央空间放射性布置使这一中央空间充满活力。该项目有三个像房子一样的体量，包括一个平面工作室（这里有丝印工作坊）展示区、一个

If cities are to continue fulfilling their essential function as places of social exchange, how should they adapt? How can public buildings become destinations that give people a reason to visit the city, and that create settings for them to meet? And how can the activity in public buildings extend to the surrounding public space, contributing to building a lively public realm? An overview of selected recent projects demonstrates how designers of public buildings are addressing these questions.

1. Drawing Activity: Public Program

The new "elective city life" is bringing new and expanded functions to the city, including facilities for cultural events, exhibitions, and sports. Large-scale events facilities include buildings such as the 62,000m² Accorhotels Arena in Paris (DVVD) with a capacity of 7,000 – 20,000 spectators, which hosts concerts and international sporting competitions.

Cultural programs encompass a greater variety of events, requiring facilities with more spatial flexibility. The expanded Palais de la Musique et des Congrès by Dietrich | Untertrifaller and Rey-Lucquet et associés in Strasbourg, for instance, contains six differently sized auditoriums and halls, fifteen conference rooms, bars, a restaurant, and office spaces. Together with the nearby Exhibition park, the conference center hosts congresses, corporate conventions, exhibitions, and cultural events.

BudCud's Jewish Culture Festival Pavilions in Kazimierz, the Jewish quarter of Kraków, also arranges programs to activate a central space. Three house-like volumes, containing a showcase for a graphic studio (where screen print workshops are held), a cafe, and a social library, open onto a covered circular stage. This central space is

当代艺术博物馆与城市规划展览馆，中国
Museum of Contemporary Art & Planning Exhibition, China

新罗马EUR区会议中心和“云”酒店，意大利
New Rome / EUR Convention Hall and Hotel "the Cloud", Italy

咖啡厅和一个社区图书馆。这三个建筑体量都面向有顶圆形舞台开放。这一中央空间有时用作举办音乐会的舞台，有时成为舞厅，有时是教室，有时成为咖啡厅的室外延伸。就是以这样的方式，这个犹太文化节日馆将不同活动浓缩在这“迷你城市一隅”，使一个被遗弃的城市广场重新焕发生机。

比上述建筑项目规模更大一些的，是蓝天组设计的深圳当代艺术博物馆与城市规划展览馆，这一设计将当代艺术博物馆和城市规划展览馆纳于同一屋檐下。这两个场馆共享一层，各自的展览也延伸至这个开放空间。两个场馆之间通过连廊连接，而且它们之间是反光的云形构件，“云”空间里有咖啡厅、书店和博物馆商店。

在城市层面上，项目混合也许是近年来交通枢纽开发项目最明显的趋势，这些交通枢纽自身也成为城市的中心。例如，由UNStudio 总体规划设计的荷兰阿纳姆中央火车站就是如此，火车站、地方汽车站和区域汽车站都汇集于此，项目还包括围绕大型中央火车站大厅而设计的两栋办公大楼、商业区和一个会议中心。火车站周围的区域还将建造总面积达 160 000m² 的办公室、商店和电影院。

2. 创建令人愉悦的场所：室内公共空间

不同的功能项目结合在一起时，它们之间室内公共空间的空间和材料质量在推动不同用户群与周边环境之间以及彼此之间

alternatively used as a concert stage, a dance floor, a classroom or an extended cafe. In this way, the pavilions condense programs into a "mini urban situation" to re-activate an abandoned city square.

On a larger scale, Coop Himmelb(l)au's design for the MOCAPE in Shenzhen combines the Museum of Contemporary Art and the Planning Exhibition under one roof. The two institutions are lifted above a continuous ground floor, and their exhibitions extend into this open space. Between these two volumes, and connected to them with a bridge, is a reflective "cloud"-shaped element containing a cafe, book shop and museum store.

At the scale of the city, program mixing is perhaps most evident in recent transportation hub developments, which are becoming urban centers in their own right. For instance, the new Transfer Terminal at Arnhem Central Station in the Netherlands, master planned by UNStudio, serves trains, local and regional busses, and includes two office towers, commercial areas, and a conference center, arranged around a large central terminal hall. The area around the station will host 160,000m² of offices, shops and a cinema complex.

2. Creating a Pleasant Place to Stay: The Public Interior

When different programs are combined, the spatial and material qualities of the public interiors between them – MOCAPE's ground floor, or Arnhem Station's terminal hall – play a fundamental role in encouraging the different user groups to engage with the surroundings and with each other.

The New Rome/EUR Convention Hall, designed by Studio Fuksas, also uses sculptural form to create a visually interesting public interior. The main public building of the convention center, the "Theca", is an elongated glass box. Inside, there is an imposing white cloud-like volume composed of a steel-ribbed structure covered with a

©Alexander Berzing

街头艺术博物馆，俄罗斯
Street Art Museum, Russia

©Luc Boegly

马赛港，法国
Marseilles Docks, France

融洽和谐方面起到至关重要的作用。当代艺术博物馆与城市规划展览馆的一层以及阿纳姆中央火车站大厅的设计就是如此。

由 Fuksas 工作室设计的新罗马 EUR 区会议中心同样使用富有雕塑感的造型创造了一个视觉上非常有趣的室内公共空间。该会议中心的主体公共建筑——"鞘"，是一个长方形玻璃结构。建筑内部，像云朵一样的白色体量非常壮观。这是一个钢肋结构，外侧用半透明的材料覆盖，内有一个可以容纳 1850 位观众的礼堂以及各种后勤服务空间。波状起伏的"云"体量三处凸出鼓起，与"鞘"结构的室内通道相连，其他地方设有开口，人们可以饱览周围的城市景色。到了晚上，"云"被点亮，成为周边地区一座富有雕塑感的明亮灯塔。

Archattacka 建筑事务所将位于圣彼得堡的一座工业建筑及其周围环境改造成了一座街头艺术博物馆。这表明，对于已经具有鲜明特色的项目环境而言，只需要进行最小限度的干预。此项目设计采用了"非设计"法，在原工业建筑之上只增加了两个盒子元素：一个用作卫生间，另一个是带有家具储存分区的展览模块。项目的外部空间由一面明亮的红墙定义，这面墙由红色的集装箱组成。原汁原味的工业元素——破旧的砖墙、高耸的圆柱形烟囱、半空中穿过院落的轨道——赋予博物馆以真实感、历史感和恒久感，同时赋予它一种非正式性，这与街头艺术主题及其他的自发创作活动非常合拍。

由 5+1AA 设计的马赛港则采用另一种不同的方法来改造一个工业园区。该项目利用丰富的材料、灯光照明和路线规划打造了一个既有工业特色又精致娴雅的公共空间。马赛港建筑位于货运港和老城中心帕尼耶之间，是一个共有六层、约 400m 长的综

translucent curtain, which contains a 1,850p auditorium and supporting services. The undulating form bulges out in three places to connect to the Theca's interior circulation galleries, and opens in other places to offer views to the surrounding city. In the evening, the cloud is lit, becoming a bright sculptural beacon for its surroundings.
Archattacka's transformation of an industrial building and its surroundings into a Street Art Museum in Saint Petersburg demonstrates that minimal intervention is needed in a setting that already has a strong character. Taking a "non-design" approach, only two box elements are added to the building: one is a washroom block, the other is an exhibition module with partitions for furniture storage. The surrounding exterior space is delineated by a wall of bright red shipping containers. The industrial elements left as they are – the worn brick walls, the tall cylindrical chimney, the raised rail tracks passing through the yard – give the museum a sense of authenticity, history and permanence, while also giving it an informal quality, suited for street art and other spontaneous uses.
5+1AA's Marseilles Docks takes a different approach to transforming an industrial complex, using rich materials, lighting, and routing to create a public space that is at once industrial and refined. The Docks are a six-story, almost 400m long complex between the cargo port and the Panier, the old city center, and were a barrier between the two areas. The new design transforms the Docks' ground floor and four inner courtyards, creating a sequence of internal squares. Each square is a miniature "world" of its own: one has a light wooden floor and walls adorned with yellow-green ceramic tiles, potted plants and red ceramic geckoes and dragon flies. Another is deep blue, with large irregularly shaped tiles covering the floor and climbing up the walls, with linear lighting elements appearing as bright cracks between the tiles. From there, visitors pass through a bright courtyard with a light stone floor and tall palm trees, before moving on to a space with large fabric lampshade-like elements hanging

©Aitor Ortiz

利马会议中心，秘鲁
Lima Convention Center, Peru

©Javier Callejas

伯明翰新街火车站，英国
Birmingham New Street Station, United Kingdom

合设施，曾是这两个区域之间的一道屏障。新的设计重新打造了马赛港的一层空间和四处内部庭院，设计了一系列内部广场。每个广场都是一个各具特色的微型"世界"：一处广场地面铺设浅色木地板，墙面用黄绿瓷砖、盆栽植物、红色的陶瓷壁虎和蜻蜓装饰；另一处广场的色调是深蓝色的，地面和墙面都用大块、形状不规则的瓷砖装饰，瓷砖之间镶嵌了带状照明灯具，闪闪发亮，看起来就像瓷砖间的裂缝。游客从这儿继续走，又可以穿过一处明亮的庭院，庭院地面由浅色石材装饰，栽种着高大的棕榈树。继续前行，来到一处由大大的灯罩似的织物设计元素装饰的空间，这些大大的灯罩悬挂于玻璃屋顶之上。除了连接这些庭院的路线之外，穿过大楼的横向道路也将港口与市中心连接起来。

3. 使活动可视化：与周围环境连为一体

公共建筑可以通过建立内部活动和周围环境之间的联系来塑造一个富有凝聚力的城市公共领域。由 IDOM 设计的利马会议中心就考虑到了这种联系。会议中心一层有一处面积为 1800m² 的空间，用可移动的隔声板隔断，拆掉隔断，就变成了一处面积达 2500m² 的城市广场。同样，该建筑的中心也是一个大型的室外空间，因不同会议厅的高度不同，这一室外空间的高度也有所不同，呈阶梯式分布。通过这个室外空间，大街上的人们可以看到建筑内部深处。

由 AZPML 事务所设计的伯明翰新街火车站通过有选择性地反映城市部分景观，使周围的公共空间充满生机与活力。该建筑用反光不锈钢立面装饰，通过弯曲、变形的形态营造了火车的动感，从而表达了火车站的动态与活力。熙熙攘攘的旅客和列车在火车站进进出出，与周围环境遥相呼应，增强了城市的活力。

from the glass roof. In addition to the continuous route passing through these courtyards, transverse paths cut across the building, linking the port with the city center.

3. Making Activity Visible: Connections to the Context

By establishing connections between the internal activity and the surrounding context, public buildings contribute to shaping a cohesive urban public realm. The Lima Convention Center, designed by IDOM, allows for such a connection with a 1,800m² room on the ground floor with movable acoustic panels, allowing it to open entirely to create a sheltered urban plaza of over 2,500m². As well, the heart of the building features a large stepped outdoor area formed by the height differences of the various convention halls. From the street, this void allows views deep into the interior of the building.

AZPML's design for the Birmingham New Street Station enlivens the surrounding public space by selectively reflecting parts of the city. The building expresses the station's dynamism with a warping, bending form that recalls the movement of trains, finished in reflective stainless steel. The crowds of passengers and trains coming in and out of the station are reflected to the surroundings, magnifying and intensifying the urban activity.

Carreño Sartori Arquitectos have also unified a previously divided city plot with their design for the Municipal Gym of Salamanca, Spain. The site was next to a soccer field and had an old gym and an abandoned pool, both of which were used as independent and private properties with blind enclosures. The new gym building instead connects to a new public square, unifying the plot and opening it to the soccer field and to the surroundings

1. Jan Gehl et al., New City Life (Copenhagen: The Danish Architectural Press, 2006), 8,12.
2. Idem, 14.
3. Ibidem.
4. Idem, 8.

©Aitor Ortiz
帕萨亚回力球场与公园，西班牙
Pelota Court & Park in Pasaia, Spain

Carreño Sartori Arquitectos 也将一处城市地块上原本四零八散的功能设施与西班牙萨拉曼卡市体育馆项目整合成一体。项目场地紧邻一个足球场，有一个老体育馆和一个废弃的游泳池，两者彼此独立、私密，四周装有围挡让人看不见里面。而新体育馆建筑与新的公共广场连为一体，既整合了整个地块，又向足球场和周围街道开放。建筑的四面都有入口，主入口位于公共广场一侧。体育馆南看台采用悬臂结构，位于主入口之上，非常引人注目，而这幢全新公共建筑就是通过体育馆南看台下的斜木墙来定义的。西立面主要为足球场辅助设施，设有一间记者室和多间公共更衣室，这样就可以功能互补，物尽其用。

同样，由 VAUMM 设计的位于巴斯克自治区吉普斯夸省的帕萨亚回力球场与公园项目围绕回力球场这一室内运动设施打造了一个全新的公共空间。作为把老工业区改造开发成居民区项目的一部分，建筑师们为该地修建了焕然一新的地形。在环形公路和莫丽纳河之间的狭长地块上，一条景观缓冲区使该地块远离车流的喧嚣。这个绿化带边上是一条新修的运动跑道，跑道两头与原来的小路相连；绿化带边上还有一个广场，有一些供老年人使用的运动器材。沿着地形往下，中间部分是回力球场建筑，配套设施还有咖啡亭和卫生间。这些设施也为邻近的公共广场提供服务。

公共建筑通过提供各种娱乐活动来应对当今"选择型"城市生活的需求，又通过使用丰富的材料、自然的光照和开阔的视野等对公共空间内部进行精心设计来鼓励用户群体之间进行社会交流。公共建筑通过将内部的活动展示给周围环境，并形成连贯一致的公共空间，成为城市的社会基础设施。因此说，公共建筑加强了城市作为人们交流往来之所这一基本功能。

streets. Access is provided from all sides of the building, with the main entrance on the public square. Definition is given to this new public space by the wood diagonal wall under the gym's south tribune, which cantilevers dramatically over the entrance. The west facade features complementary uses for the soccer field – a reporter's room and shared dressing rooms – allowing them to function together.
Likewise, VAUMM's Pelota Court and Park in the Basque province of Gipuzkoa creates new public space around a Pelota court indoor sport facility. As part of the redevelopment of the previously industrial region into a residential area, the architects built an entirely new topography for the location. In the strip of land between a ring road and the Molinao river, a landscaped buffer zone shields the area from traffic. Next to this green strip is a new running trail that connects at both ends with existing paths, and a square with sport sets for elderly people. Lower down, in the intermediate area of the topography, is the Pelota court building, with an integrated "kiosk" café and toilet that also serve the adjacent public square.

Public buildings respond to the demands of today's "elective" city life by offering a wide variety of recreational activities, and encourage social exchange between user groups by arranging programs around well-designed public interiors with rich materialization, natural light, and views. By making their activity visible to the surroundings, and by contributing to making a coherent and continuous public ground, public buildings serve as the functional anchors of an urban social infrastructure. In this way, they strengthen the city's essential function as a place of communication and exchange. Isabel Potworowski

新罗马EUR区会议中心和“云”酒店

Studio Fuksas

©Moreno Maggi (courtesy of the architect)

新罗马 EUR 区会议中心和"云"酒店由意大利建筑事务所 Fuksas 设计完成，是罗马五十多年来最大的建筑综合体。此建筑具有抗震性，共耗资 2.39 亿欧元，从规划设计到落成共历时 18 年，于 2016 年 10 月对公众开放。综合体建筑的公共空间共 55 000m²，包括礼堂、展厅和酒店。该会议中心通过自身业务和旅游业，有望给罗马市每年带来 3 亿至 4 亿欧元的收入。综合体建筑位于罗马市中心的南部，这儿是 EUR 区商务区，外观为简单的正交线组合，遵循了周围 20 世纪 30 年代理性主义的建筑风格。会议中心周边的空间将作为两个公共的广场，将综合体建筑与周边环境紧密联系起来，为市民提供新的休闲和户外活动场所，也为罗马这一繁华地带创造了一个新的聚集场所。

新罗马 EUR 区会议中心和"云"酒店的设计由三大不同的建筑理念组成：地下层，"鞘"与"云"，"刀锋"。

地下层直通哥伦布大道，通过楼梯可以到达该建筑的主门厅和问讯处。穿过这片区域，宽阔的通道将人们带入宽敞的可容纳 6000 人的会议大厅和展厅。

"鞘"这一建筑理念体现在会议厅和酒店那令人叹为观止的外立面上，它由金属、玻璃和钢筋混凝土组合而成。在建筑内部，面积达 7800m² 的新建公共空间可举办公共和私人性质会议、展览和一些大型活动。"云"漂浮在"鞘"结构内，二者相互作用，对整个建筑综合体来说非常重要，象征着罗马城和会议中心之间的联系。"云"是一个独立的茧状结构，"鞘"在横向的几个点上对其起着支撑作用。"云"的表面积达 15 000m²，其材质为高科技玻璃纤维膜和阻燃硅胶。"云"位于建筑的中心处，人们可以通过连接着"云"和"鞘"的主要通道直接到达"云"的内部。这条通道被称作"论坛"。"云"结构内部共有五层，人们可通过电梯和通道进入能容纳 1800 人的礼堂之中。为了确保"云"系统不会对整座建筑的其他系统造成声音干扰，礼堂四周墙面采用樱桃木面板包裹。

最后一个建筑理念是"刀锋"，一栋可以自我管理的大楼，共 17 层，是一个拥有 439 间客房的酒店，为前来罗马城和会议中心的客人提供住宿。"刀锋"建筑面积超过 18 000m²，内部除了一般客房外，还有 7 间精品套房、一个 SPA 养生会所和一家餐厅。

建造该建筑综合体共使用了 37 000t 钢材，其重量相当于四个半埃菲尔铁塔的重量。此外，这栋建筑的外立面以及室内设计共消耗了 58 000m² 的玻璃，足以覆盖十个足球场。

会议中心全面采用抗震设计——其垂直结构的刚度能够抵抗大大小小的地震波。

此外，建筑保温材料的水平刚度能够抵挡小型地震，当遇到较强地震时，其低刚度可以减缓大幅度的震动。

该建筑设计集中体现了生态环保的理念：中央空调由一个可逆的热泵带动，在提高能效的同时降低了电力消耗。建筑内部自然通风系统的设计也是恰到好处，将附近 EUR 区凉爽的湖水汲取并过滤到通风系统中。屋顶设有由玻璃和硅晶片构成的光伏电池板，不仅可以提供能量，而且能够减少太阳辐射，避免建筑过热。当新罗马会议中心和"云"酒店全面投入运营后，其基础用电将由废热发电、地热发电和光伏发电共同提供。这三个发电系统相辅相成，相互依存，如果出现任何技术故障，也能够确保整座建筑正常运行。

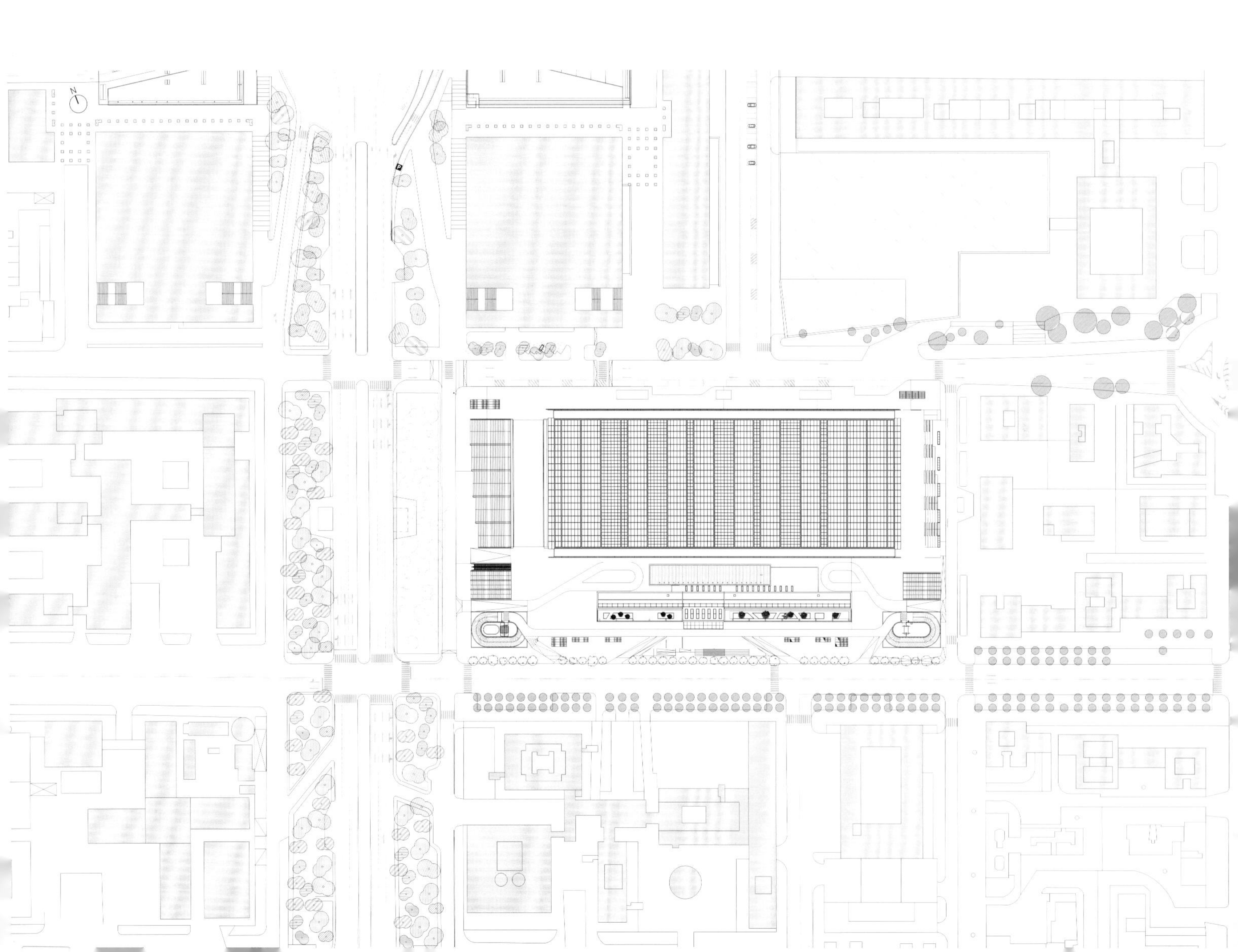

会议中心的生态设计特色还包括一个雨水收集系统，雨水可以通过外立面的面板过滤到蓄水池中，然后根据需要通过水泵压入建筑内水系统，供会议中心使用。

New Rome / EUR Convention Hall and Hotel "the Cloud"

Italian architecture practice Studio Fuksas has completed the largest building in Rome in over 50 years. Opening to the public in October 2016, the New Rome/EUR Convention Hall and Hotel "the Cloud" is a €239 million earthquake proof complex that has taken 18 years of planning and construction. It will host auditoriums, exhibition spaces and a hotel – amassing 55,000 square meters in new public space. Through both trade and tourism, the convention center is expected to bring in between €300~400 million annually to the city of Rome. Located south of the city's core, in the business district of EUR, the complex follows the simple orthogonal lines of the surrounding 1930s rationalist architecture. The spaces surrounding the center will serve as two public squares. Integral to the new complex and the neighbourhood, these new spaces will

provide citizens with places for various leisure and outdoor activities, offering a new meeting area in this busy part of Rome.

The New Rome/EUR Convention Hall and Hotel "the Cloud" comprises three distinct architectural concepts: the basement, the "Theca" and "Cloud", and the "Blade".

The basement is accessed on Viale Cristoforo Colombo, via a staircase that leads into the building's main foyer and information point. Past this area, a large concourse feeds into an expansive congress and exhibition hall that can host up to 6,000 people.

The "Theca" is the stunning outershell and facade of the convention Hall and Hotel, which has been made from a combination of metal, glass and reinforced concrete. Inside the building, 7,800 square meters of new public space will play host to public and private conferences, exhibitions and large-scale events. Suspended inside the "Theca" is the "Cloud" – the interplay between these two spaces is essential to the complex – symbolising the connection between the city of Rome and the convention center. The "Cloud" is an

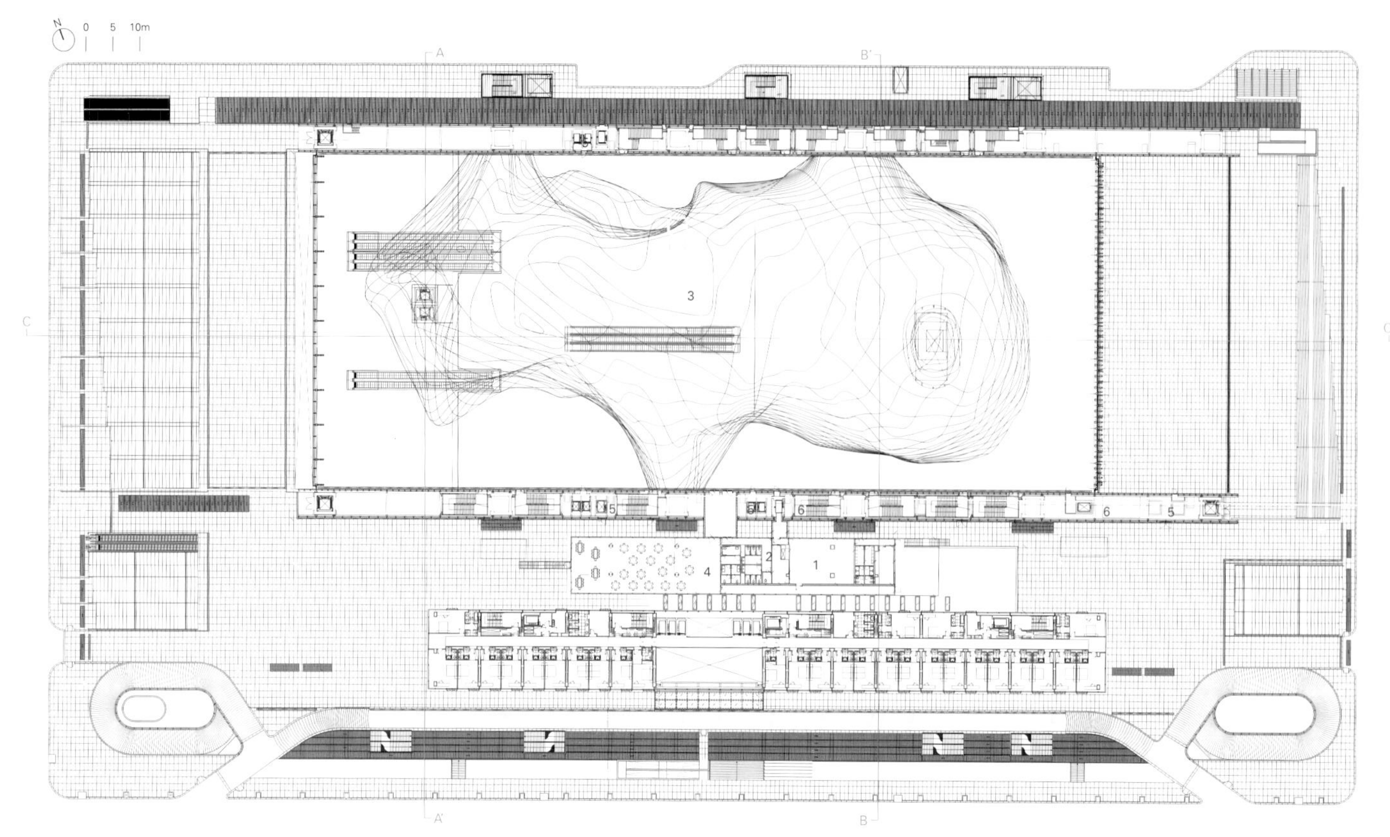

1. 厨房 2. 设备间 3. 论坛 4. 餐厅/酒吧 5. 货梯 6. 客体
1. kitchen 2. services 3. forum 4. restaurant/bar 5. lifts 6. elevators
一层 ground floor

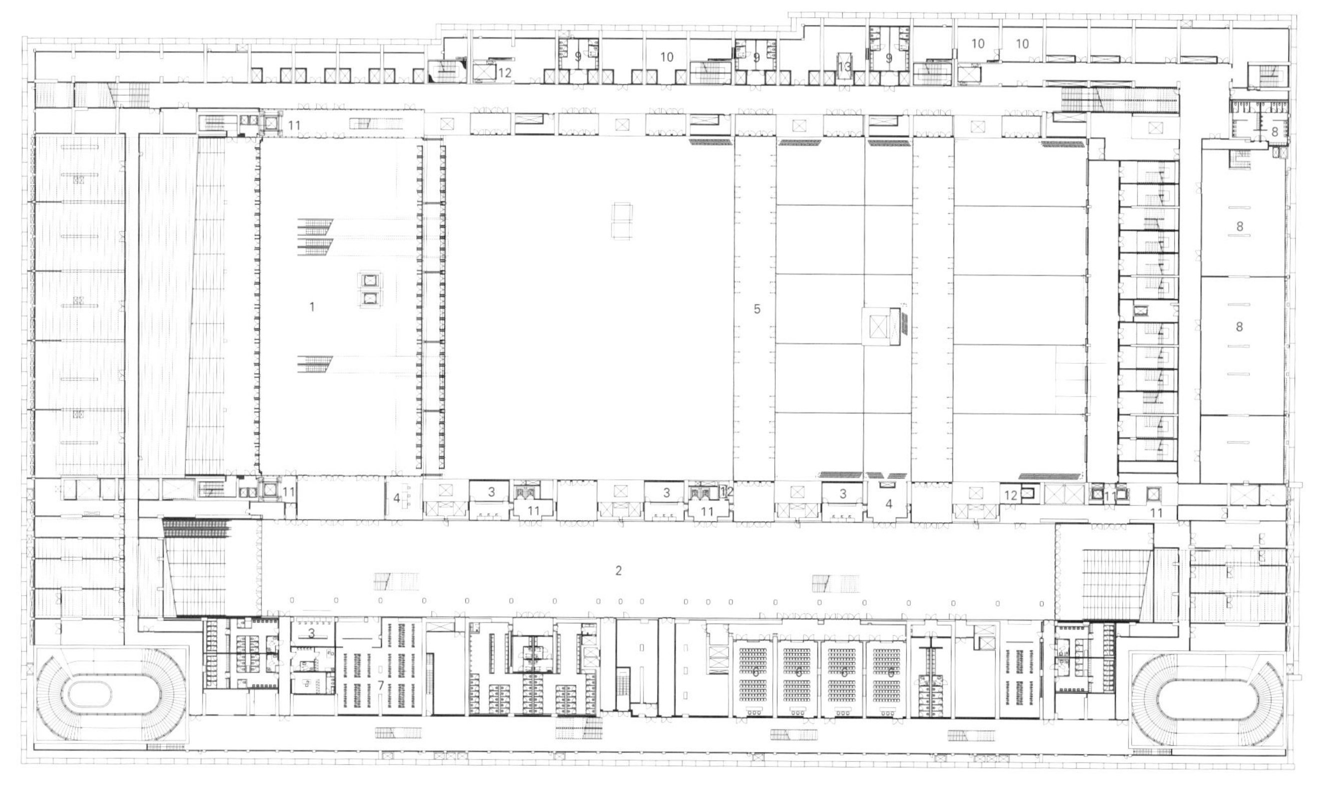

1. 门厅 2. 通道 3. 四间前台办公室 4. 问讯处 5. 多功能大厅 6. 四间会议室
7.衣帽间 8. 厨房 9. 设备间 10. 三间储藏室 11. 货梯 12. 客梯 13. 平台货梯
1. foyer 2. concourse 3. four front desk office 4. info point 5. multipurpose hall 6. four meeting rooms
7. cloakroom 8. kitchen 9. services 10. three storerooms 11. lifts 12. elevators 13. platform lift
地下一层 first floor below ground

项目名称：New Rome/EUR convention Hall and Hotel "the Cloud" / 地点：At the corner of via C. Colombo and viale Asia, Rome-Eur, Italy / 建筑师：Massimiliano, Doriana Fuksas_Fuksas Design / 室内设计：Fuksas Design / 总承包商：Società Italiana per Condotte d'Acqua SpA / 工程（规划）：A. I. Engineering
结构工程师：Studio Majowiecki; Studio Sarti / 安全：Studio Sarti / 声学设计：XU-Acoustique, A.I. Engineering / 照明顾问：Speirs & Major Associates

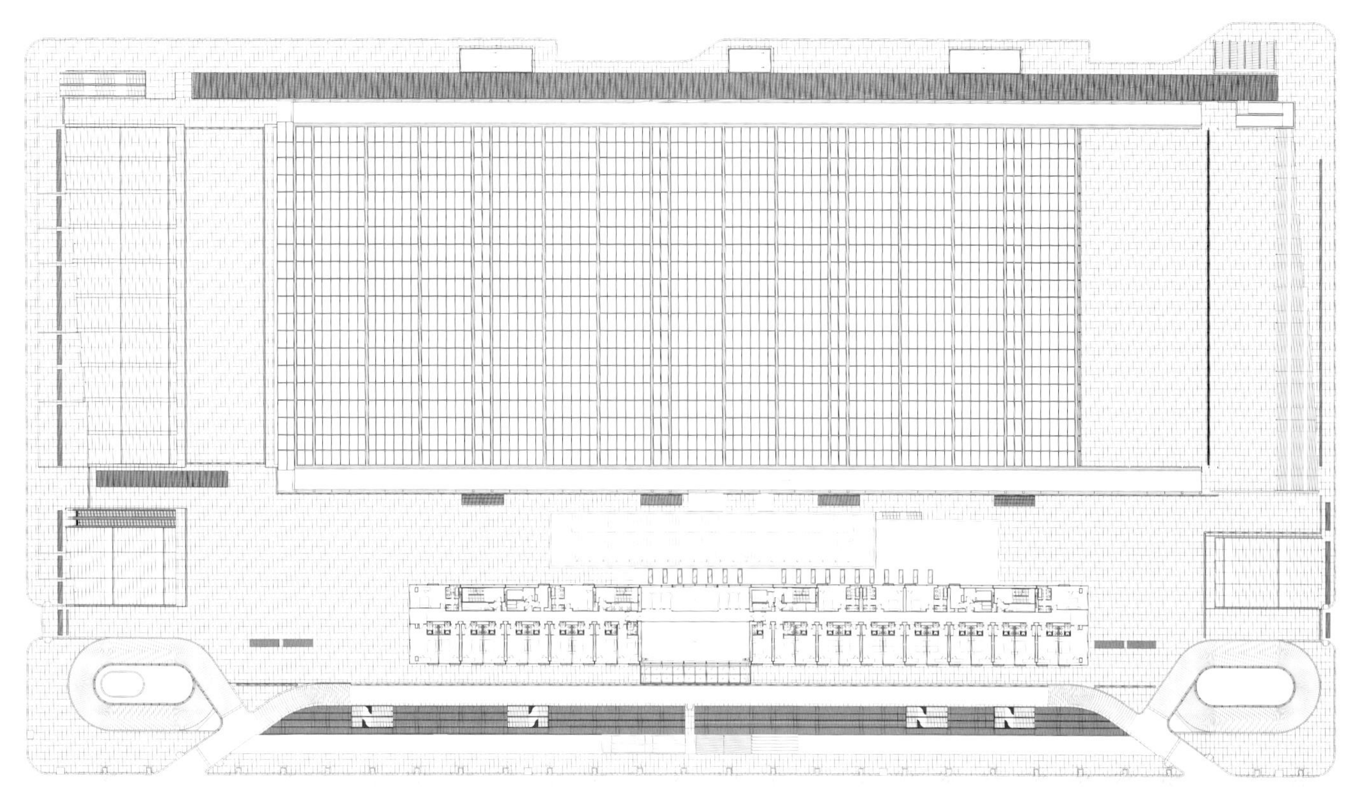

九层 eighth floor

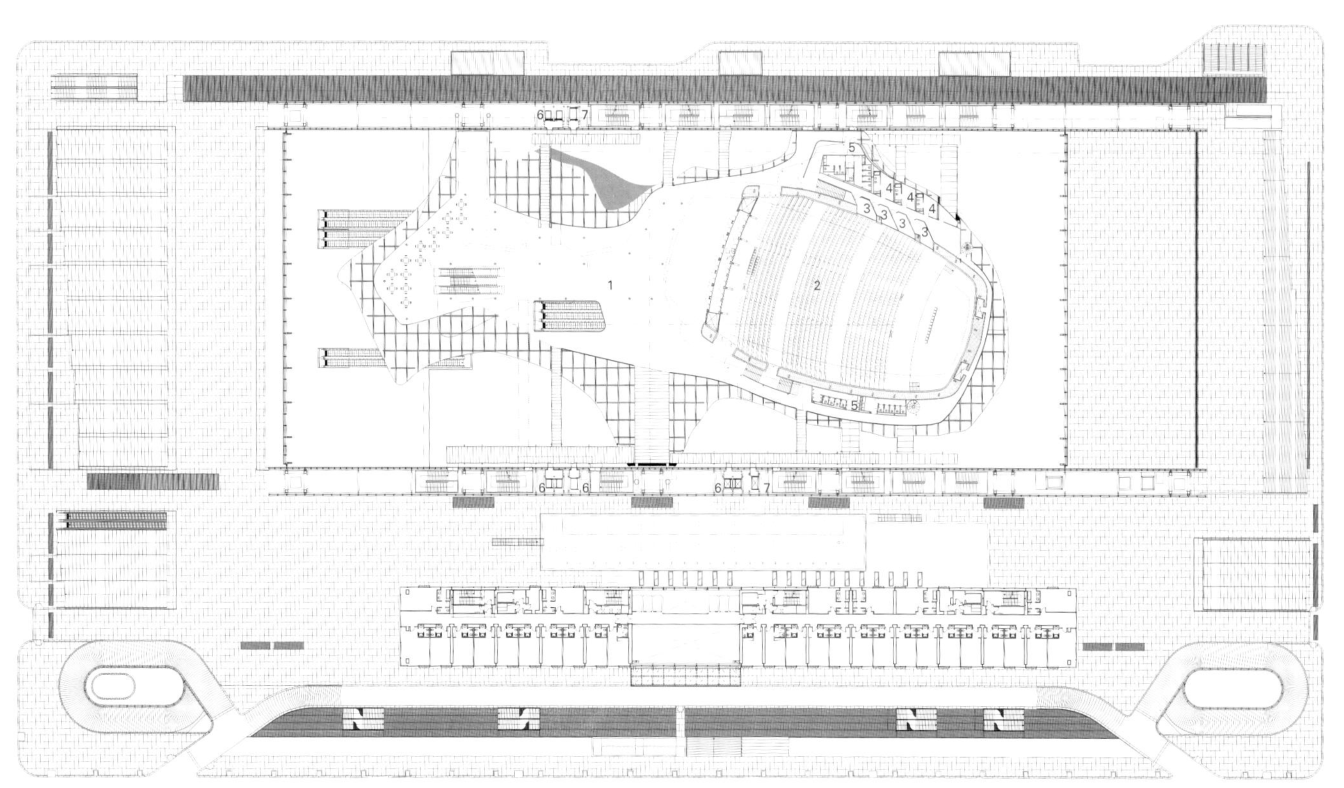

1. 门厅 2. 大厅 3. 四间翻译室 4. 三间更衣室和设备间 5. 设备间 6. 货梯 7. 客梯
1. foyer 2. hall 3. four translation rooms
4. three dressing rooms and services 5. services 6. lifts 7. elevators
四层 third floor

客户：Eur SpA / 摄影师：©Leonardo Finotti (courtesy of the architect) (except as noted) / 用地面积：27,250m² / 建筑面积：55,000m² / 造价：€ 239,000,000 / 结构：metallic and reinforced concrete_Theca, Blade (hotel); metallic_Cloud / 规模：height 39m (from the underground level 48m) x width 70m x length 175m / 竞赛时间：1998 / 施工时间：2008.3－2016 / 竣工时间：2016

independent cocoon-like structure that is covered in 15,000 square meters of highly advanced membrane fiber glass and flame-retardant silicone and is supported laterally at points by the "Theca". It lies at the heart of the complex and is accessed by the "Forum" – an artery walkway that fuses the two structures together. Inside the "Cloud", five levels supported by escalators and walkways lead to a 1,800 capacity auditorium. In order to ensure that the "Cloud" system does not interfere with the rest of the complex, the auditorium is clad in wooden cherry panels.

The final architectural concept is the "Blade" – an autonomous building split into 17 floors and containing a new 439-room hotel built to provide accommodation to visitors to the center and the city of Rome. Spread over 18,000 square meters, the "Blade" will also include seven boutique suites, a spa and a restaurant.

The building has been constructed from 37,000 tons of steel- the equivalent weight of four and a half Eiffel Towers. Additionally, 58,000 meters of glass has been used for the center's exterior and interior design, which is enough to cover the surface of 10 football pitches.

The center is fully earthquake-proofed – the stiffness of its vertical structure is able to withstand both small and large seismic waves.

©Moreno Maggi (courtesy of the architect)

In addition, the building's insulators have a horizontal rigidity, which works against the movements of small earthquakes, whilst their low rigidity enables large oscillations with low accelerations during more violent tremors.

An eco-friendly approach underscores the design of the center, with integrated airconditioning that will be carried out by a reversible heat pump. This system is capable of achieving high energy performances whilst reducing electricity consumption. A natural ventilation system is also in place – with the cool water of the nearby EUR lake extracted and filtered into the system. The roof's photovoltaic panels (glass and silicon wafer) help to produce energy and protects the building from overheating through the mitigation of solar radiation. When fully operational, the basic power load of the New Rome/EUR convention Hall and Hotel "the Cloud" will be supplied by the power station of cogeneration as well as any power generated by the buildings' geothermal and photovoltaic network. The mutual interdependence of these systems ensures that the complex is able to function in any instances of a technical failure.

The center's eco features also comprise a rain water harvesting system, where exterior panels collect rainwater and filter it into a storage tank. The water can then be pumped, on demand, from the tank to the internal water system.

©Moreno Maggi (courtesy of the architect)

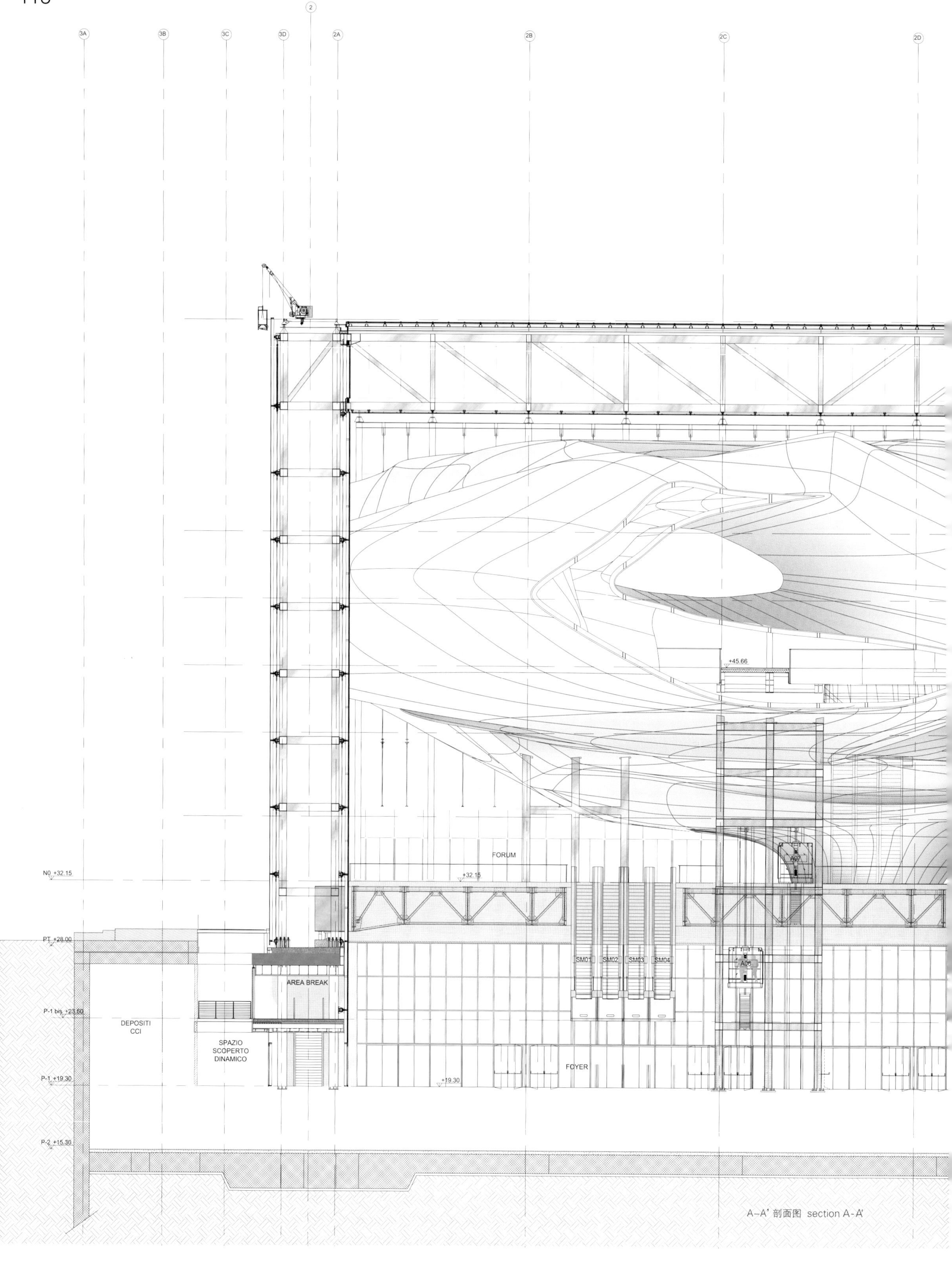

A–A′ 剖面图 section A-A′

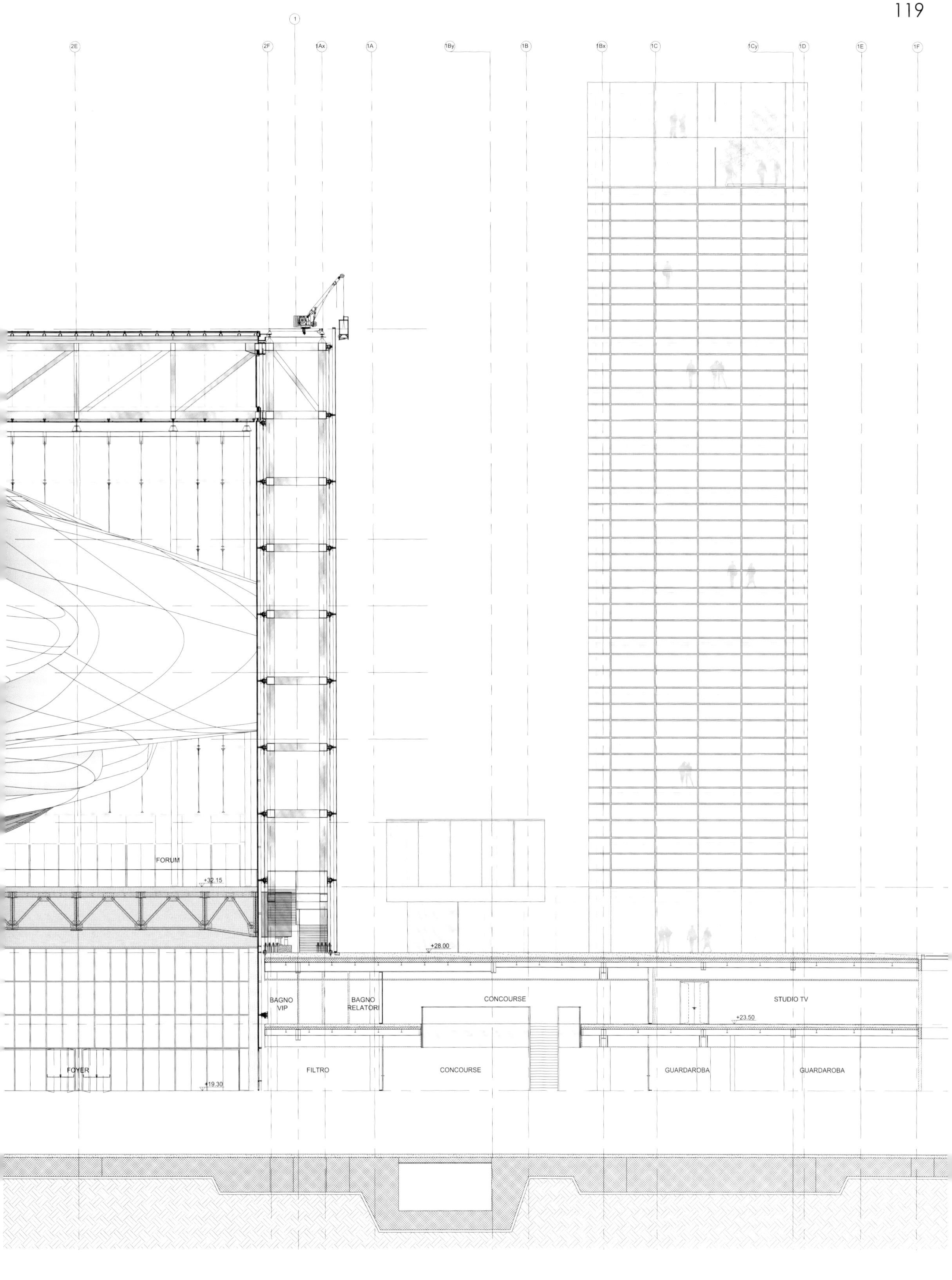

2E 2F 1 1Ax 1A 1By 1B 1Bx 1C 1Cy 1D 1E 1F

FORUM

+32.15

+28.00

BAGNO VIP

BAGNO RELATORI

CONCOURSE

STUDIO TV

+23.50

FOYER

+19.30

FILTRO

CONCOURSE

GUARDAROBA

GUARDAROBA

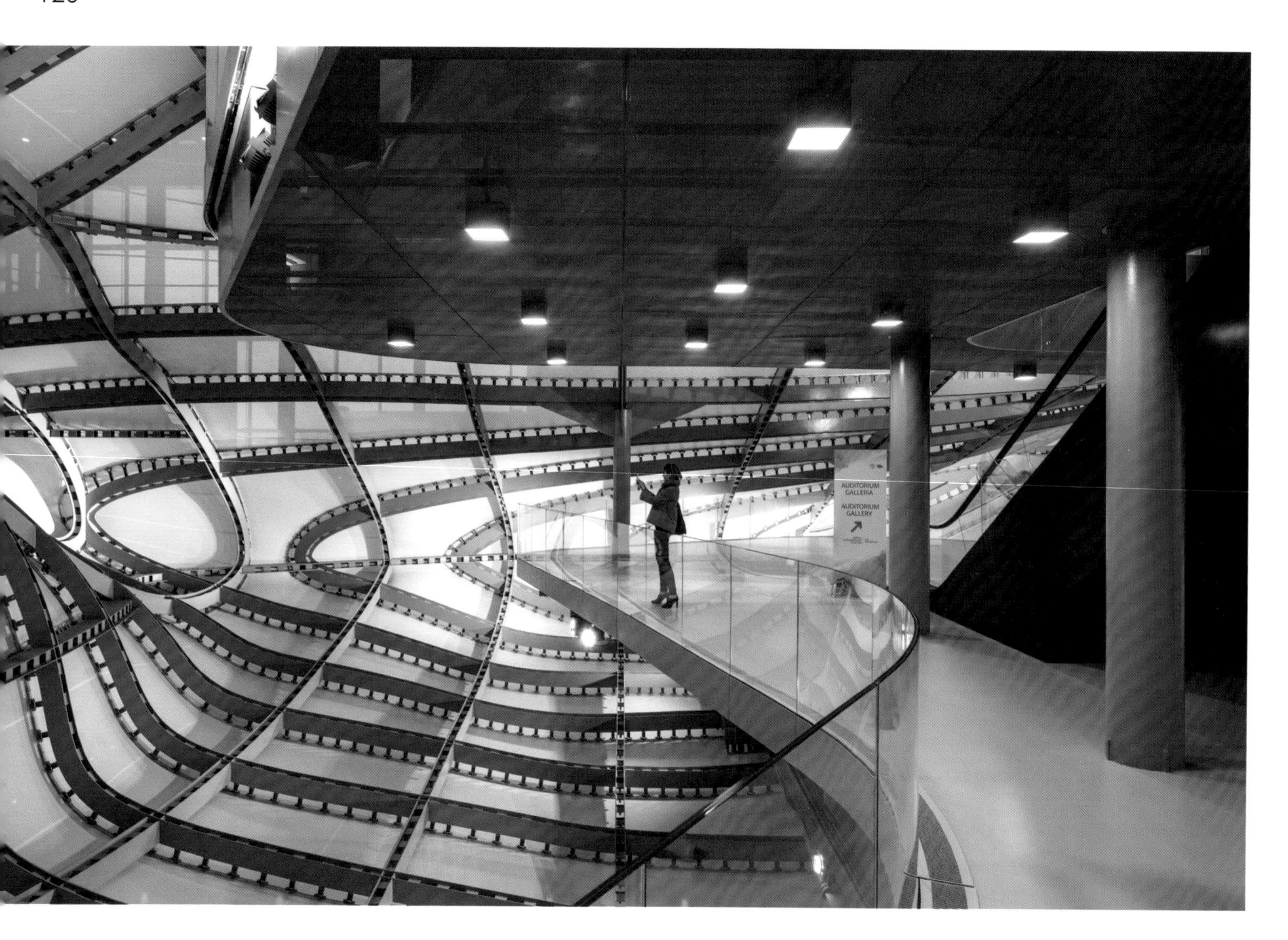
AUDITORIUM
GALLERIA
AUDITORIUM
GALLERY

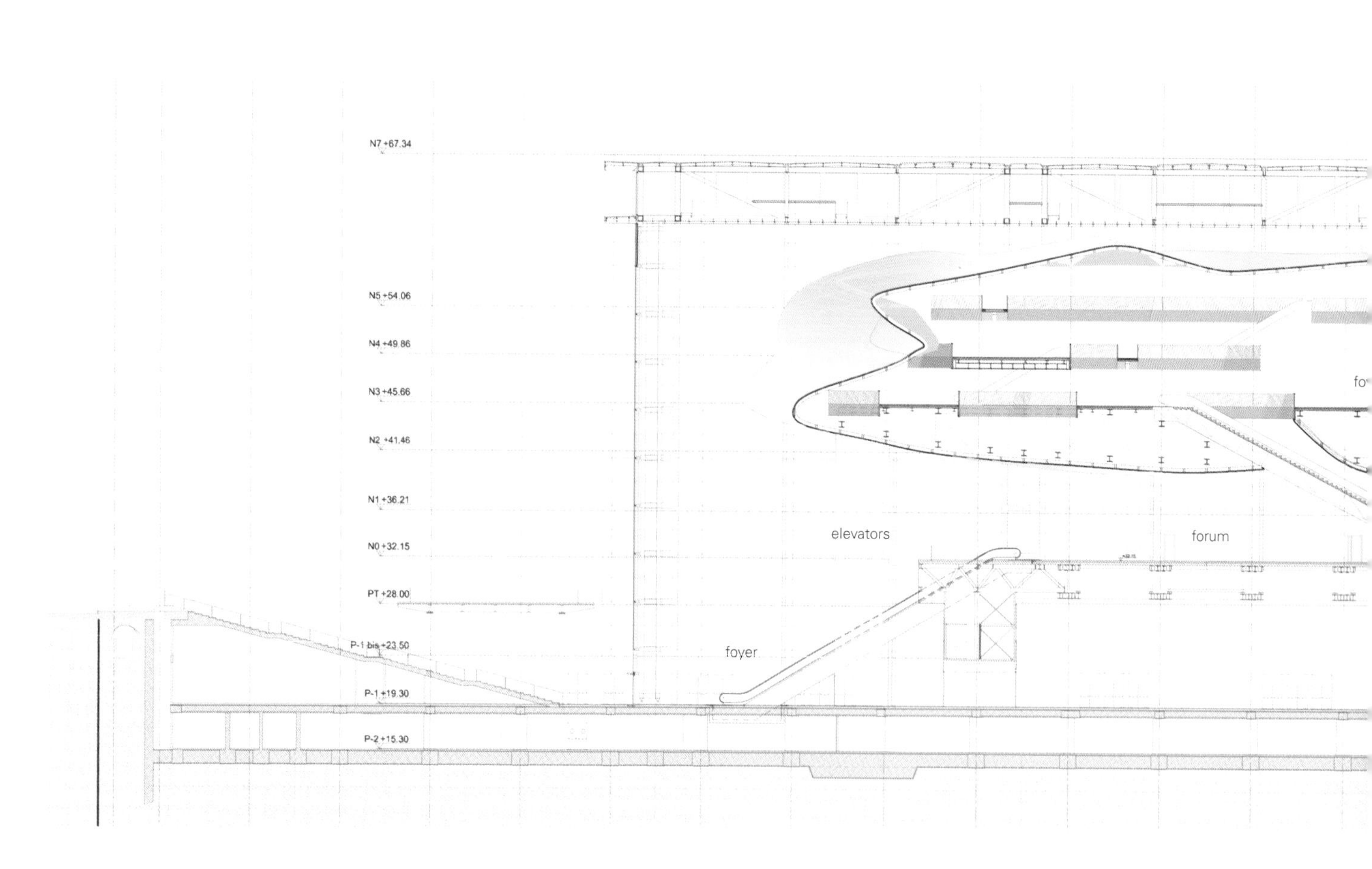
N7 +67.34
N5 +54.06
N4 +49.86
N3 +45.66
N2 +41.46
N1 +36.21
N0 +32.15
PT +28.00
P-1 bis +23.50
P-1 +19.30
P-2 +15.30
elevators
forum
foyer

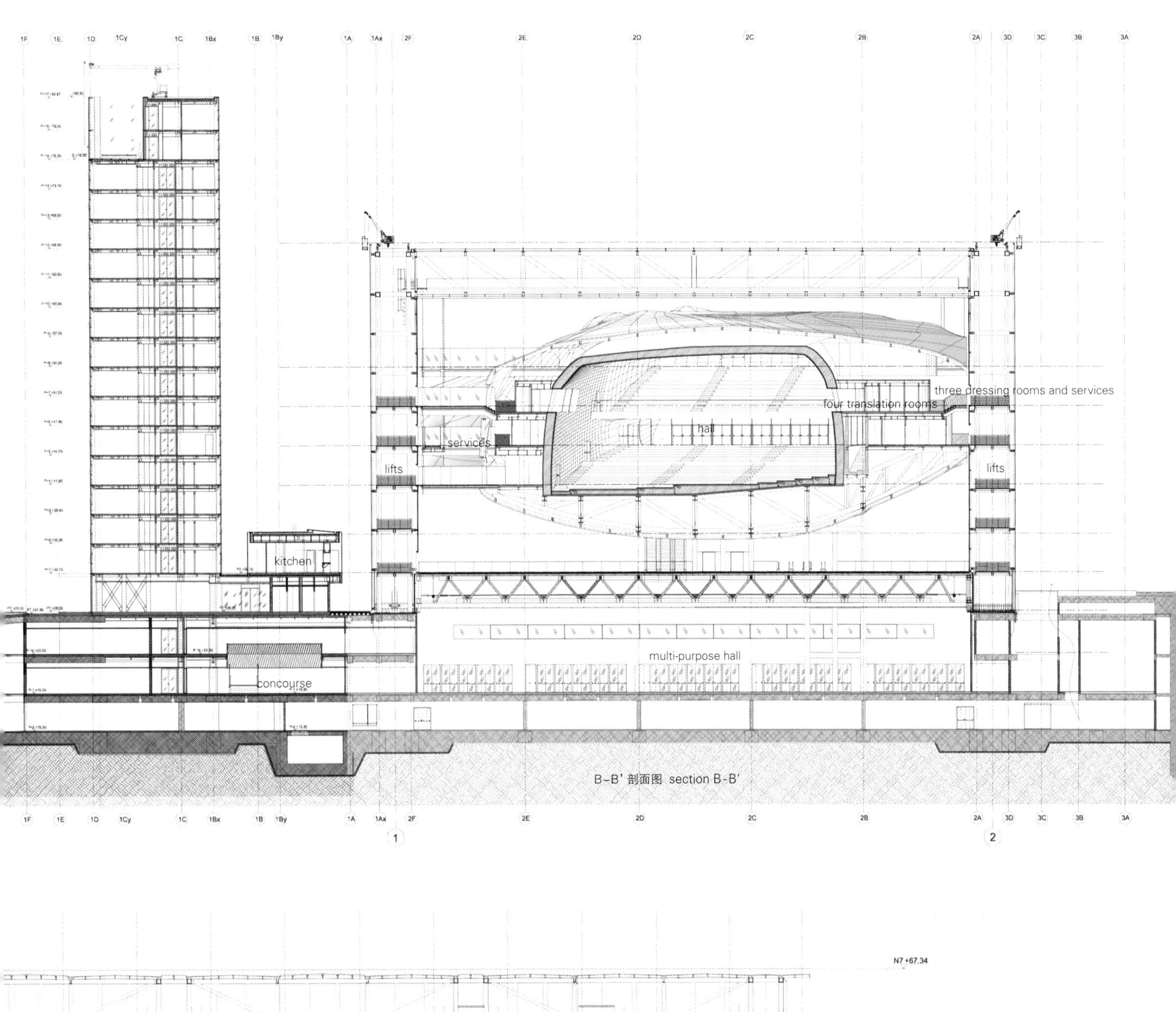

B–B' 剖面图 section B-B'

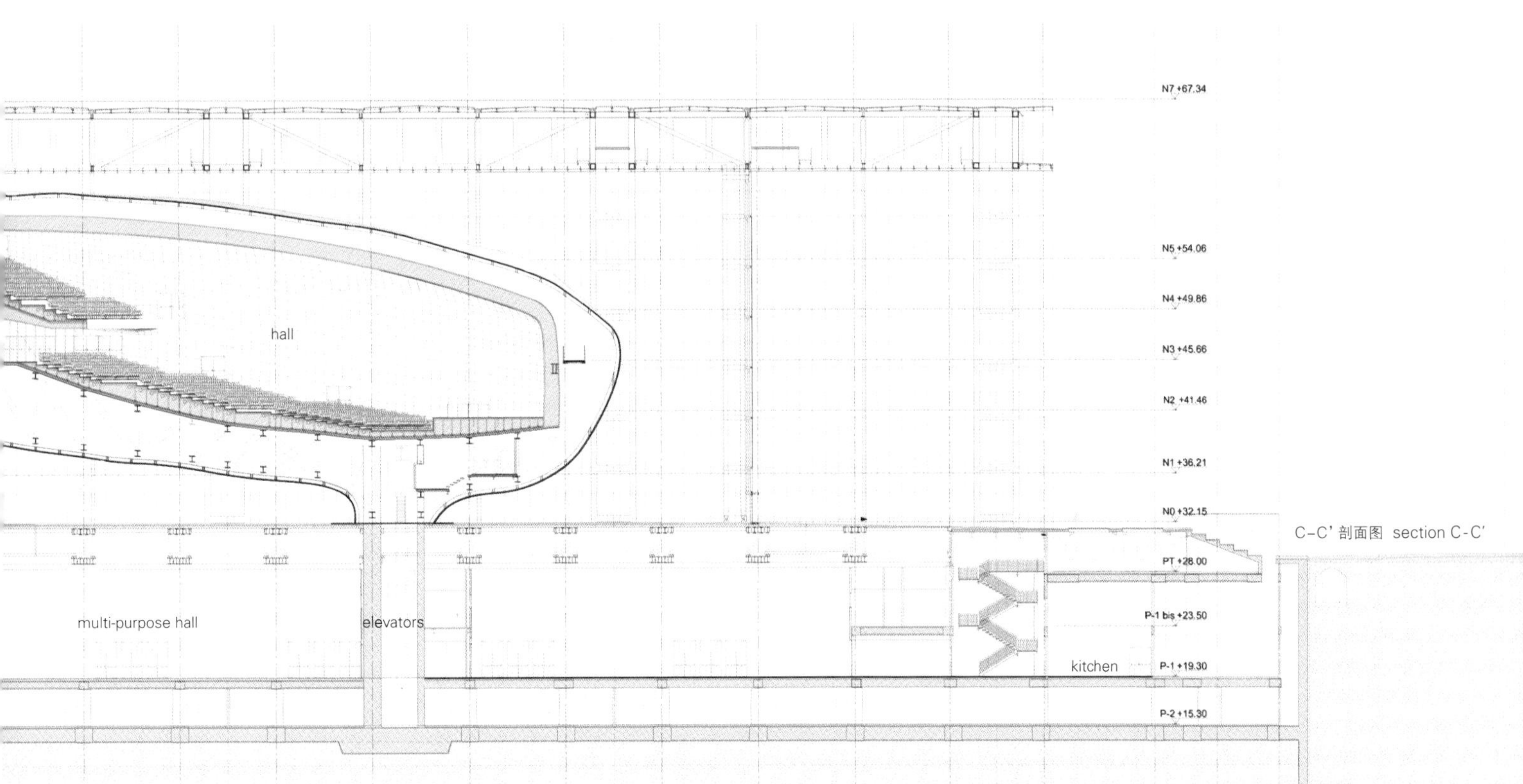

C–C' 剖面图 section C-C'

©Moreno Maggi (courtesy of the architect)

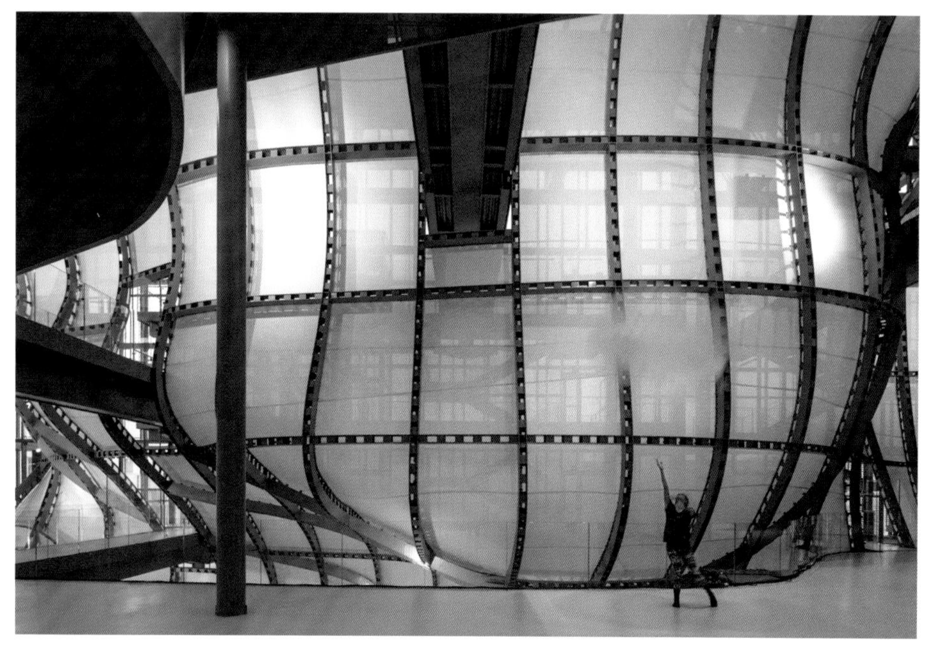

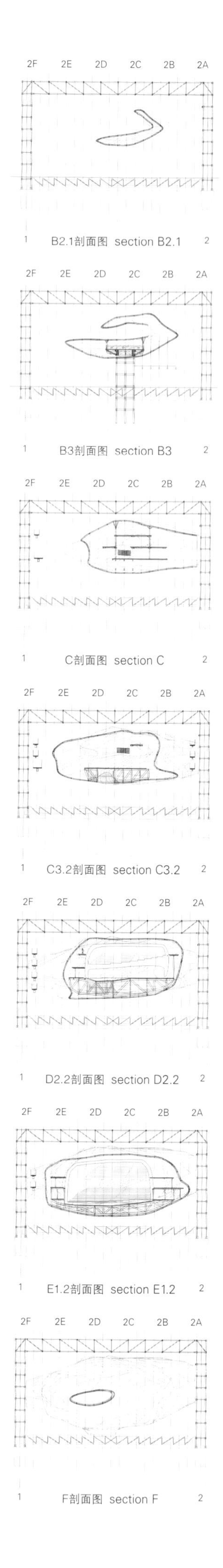

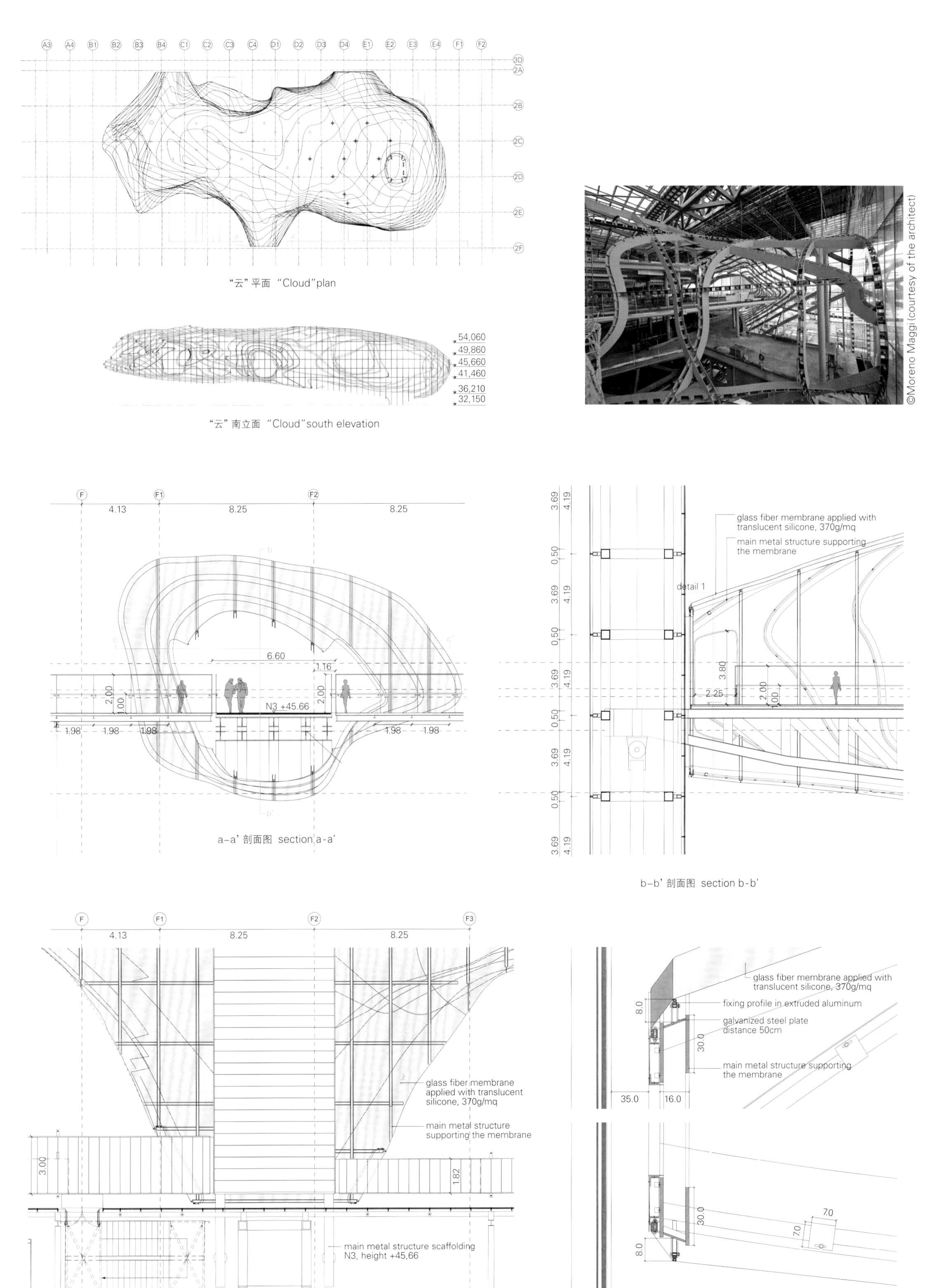

"云"平面 "Cloud"plan

"云"南立面 "Cloud"south elevation

©Moreno Maggi (courtesy of the architect)

a-a' 剖面图 section a-a'

b-b' 剖面图 section b-b'

c-c' 剖面图 section c-c'

详图1 detail 1

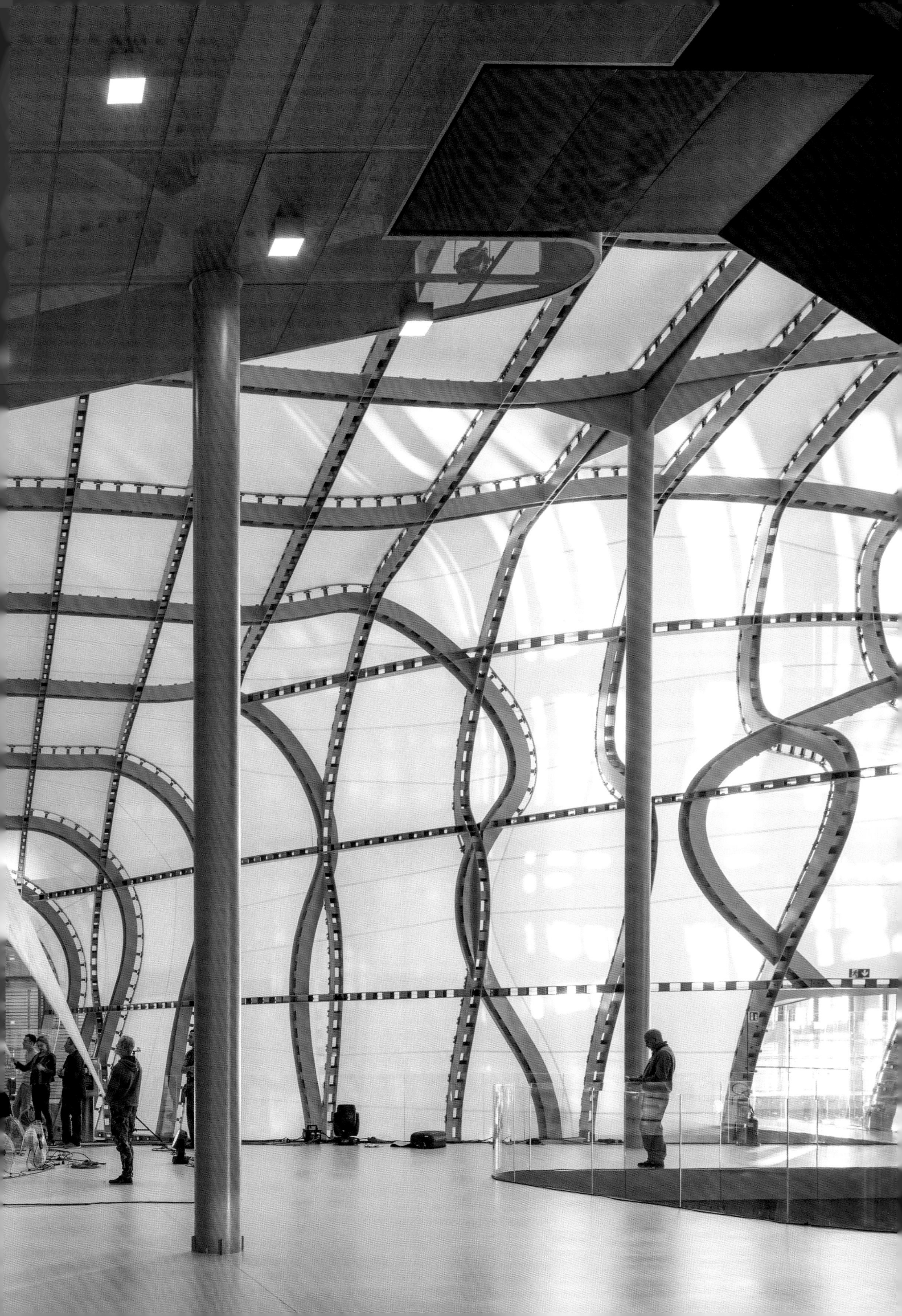

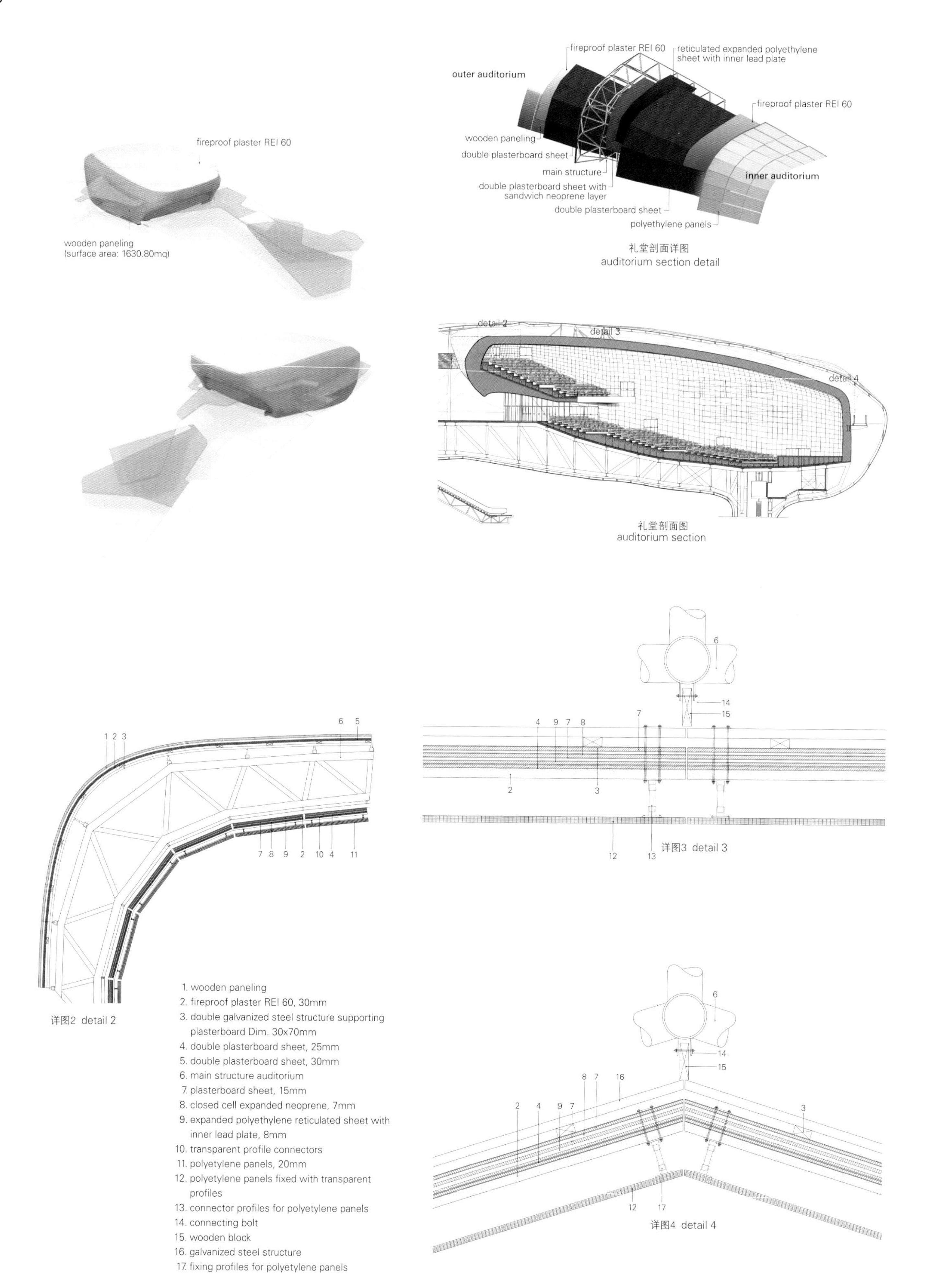
fireproof plaster REI 60
wooden paneling
(surface area: 1630.80mq)
outer auditorium
fireproof plaster REI 60
reticulated expanded polyethylene sheet with inner lead plate
fireproof plaster REI 60
wooden paneling
double plasterboard sheet
main structure
double plasterboard sheet with sandwich neoprene layer
double plasterboard sheet
inner auditorium
polyethylene panels
detail 2
detail 3
detail 4
1. wooden paneling
2. fireproof plaster REI 60, 30mm
3. double galvanized steel structure supporting plasterboard Dim. 30x70mm
4. double plasterboard sheet, 25mm
5. double plasterboard sheet, 30mm
6. main structure auditorium
7. plasterboard sheet, 15mm
8. closed cell expanded neoprene, 7mm
9. expanded polyethylene reticulated sheet with inner lead plate, 8mm
10. transparent profile connectors
11. polyetylene panels, 20mm
12. polyetylene panels fixed with transparent profiles
13. connector profiles for polyetylene panels
14. connecting bolt
15. wooden block
16. galvanized steel structure
17. fixing profiles for polyetylene panels

礼堂剖面详图
auditorium section detail

礼堂剖面图
auditorium section

详图2 detail 2

详图3 detail 3

详图4 detail 4

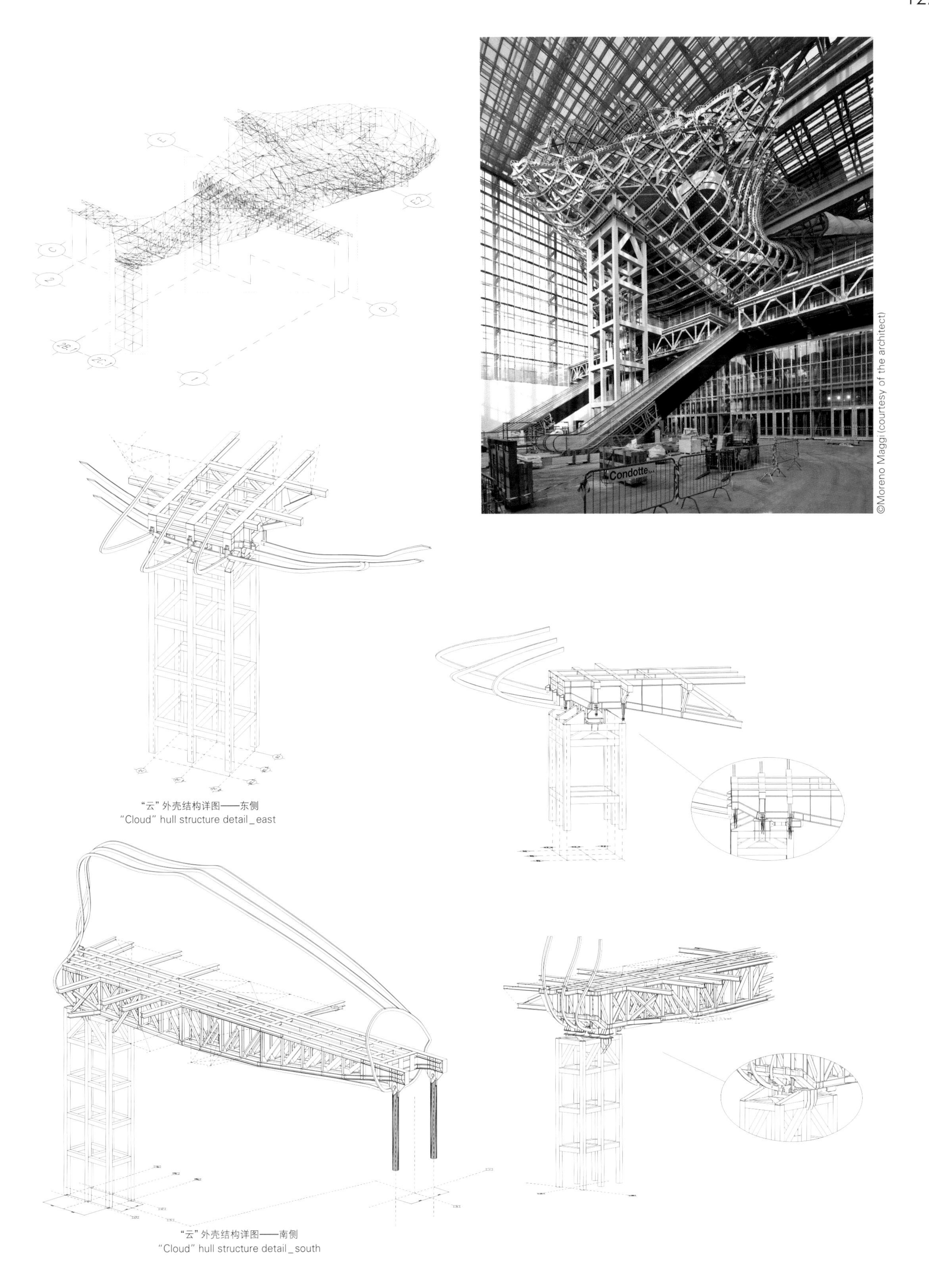

©Moreno Maggi (courtesy of the architect)

"云" 外壳结构详图——东侧
"Cloud" hull structure detail_east

"云" 外壳结构详图——南侧
"Cloud" hull structure detail_south

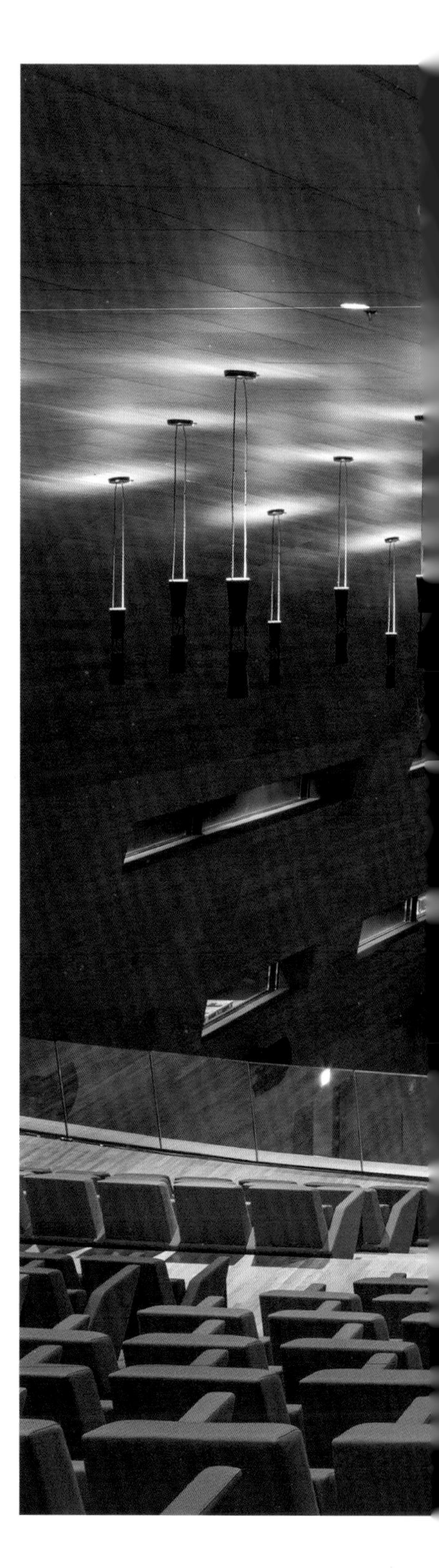

马赛港

5+1AA Alfonso Femia Gianluca Peluffo

PRINTEMPS

GAIUS IULIUS CAESAR
ALBERT LONDRES
Une touche naturelle
simply food

CHARLES DE BROSSES
S DOCK

2009年5+1AA建筑事务所赢得重建历史建筑马赛港的设计竞赛。21 000m² 的地下空间以及地上一层被翻新，主要用作经营当地工艺品的商家店铺。马赛港建筑及场地位于城市与海洋之间，沿着独特地贯穿整个周边地区的历史轴线而设。因此，建筑师需要使建筑与自然环境融为一体。而对马赛港的重新定义，将为城市与海洋之间的联系创造一个更加通透、透明的港口景象。无论是北面当代城市、南面历史名城，还是东面现代都市和西面恢复生机的海港这些城市定位，其城市与海洋之间的联系都将通过马赛港得到加强。

马赛港建筑经过重新设计打造，两端分别与两大广场直通相连：一个是有着历史文化底蕴的"Joliette 广场"，另一个是新兴的"地中海广场"。因此，马赛港重建后的空间沿着纵向轴线，与整个城市连为一体。通过这种方式，此项目能够将该地区与马赛的其他地方紧密地联系起来。马赛港建筑的设计与重建也能够促进其与周边地区的交流。这样，一个有着浓郁的地中海风格、独一无二的社交平台便建成了。

为了打开原建筑封闭的空间系统，建筑师对原建筑的基础结构进行了重新安排与设计，利用穿透性和不同层次的透明度，使建筑与海洋始终保持着密切的联系。该设计重新定义了四个庭院。这四个庭院设计充满商业色彩、艺术色彩和文化色彩，从建筑内部一直向两个广场和毗邻的两条街道延伸，在横向通道上为人们创造了四个中间休息处。

建筑师通过大胆使用不同的材料、令人意想不到的风景以及恰到好处的植物，很好地表现了重新打造的马赛港这一标志性建筑所着重强调的三大主题：港口、村落和市场。

与建筑本身一样，所使用的材料和空间元素都具有一定的象征意义与纪念意义。它们被重新诠释成亲密的情感和奇迹。不同的日常生活变换、交融，使四个庭院中的每一个都呈现出独一无二的氛围。此外，基于对季节的精确计算，建筑师在空间和其使用者之间建立了一个新的时间关系。

四个庭院的陶瓷材料折射出时间与自然的关系。随着太阳位置的移动，光线也跟着变幻，五彩缤纷的颜色和香味溢满庭院，编织出自然与人工技巧、感知与存在、情感与理性之间的对话。与此同时，日常生活的忙碌与喧嚣被屏风和挂帘过滤、隔离开来。各种不同的元素共同构建了这样的空间，一个不会对这座城市和她的人民漠然处之的空间。

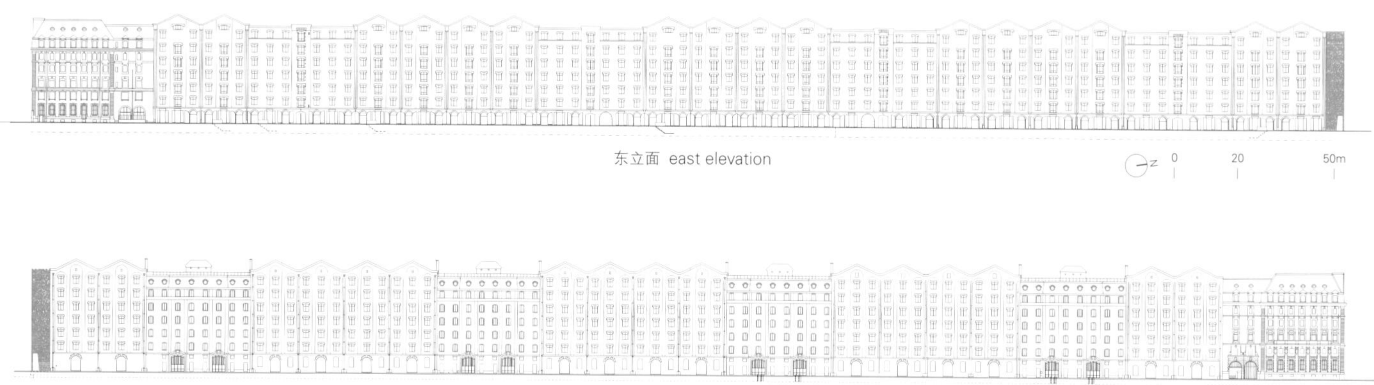

东立面 east elevation

西立面 west elevation

Marseilles Docks

In 2009, 5+1AA won the competition for the redevelopment of the historical building, Les Docks de Marseilles. The 21,000 sqm of ground floor and basement space was refurbished to host local shops mainly dealing with the craftwork of Marseilles and its regions. Located between the city and the sea, the building and its site are laid along the historical axis that distinctively runs throughout the surrounding territory. This allowed the architects to design a place that integrates its built and natural environments. In result, the redefined space of Les Docks created a new landscape permeable and porous towards the urban and the maritime. This relationship extends and develops to the contemporary city of the north, the historical city of the south, the modern city of the east, and the recovered harbor of the west.

Thus, the recreated space with its longitudinal axis continues on to the city by respectively opening up to the two squares at both ends of the building: the historical "Place de la Joliette" and the emerging "Place de la Méditerranée". In this way, the project provides spaces that actively interact with different parts of Marseilles. The building is also reorganized and reinterpreted to communicate with its surroundings. A unique sociality is created in the new places strongly characterized by its Mediterranean setting.

An intervention on the existing building base was done in order to open a closed system of spaces. Articulated with penetrations and different layers of transparencies, the cre-

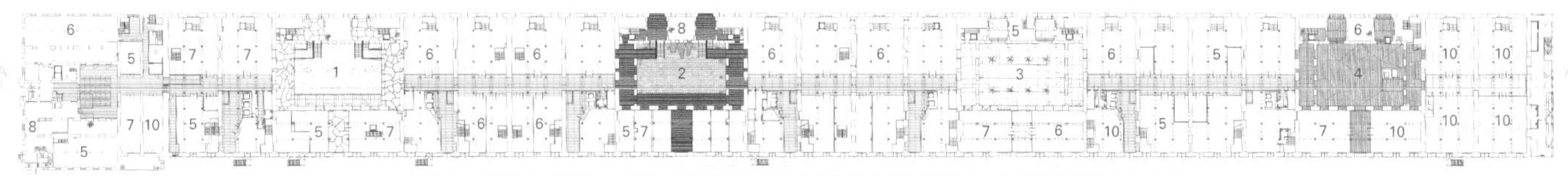

1. 庭院A 2. 庭院B 3. 庭院C 4. 庭院D 5. 设备间 6. 商店 7. 餐厅 8. 咖啡厅/酒吧 9. 文化馆/洛伊斯酒店 10. 市场
1. courtyard A 2. courtyard B 3. courtyard C 4. courtyard D 5. services 6. shop 7. restaurant 8. cafe / bar 9. culture / Loisir 10. market
一层上部 higher ground floor

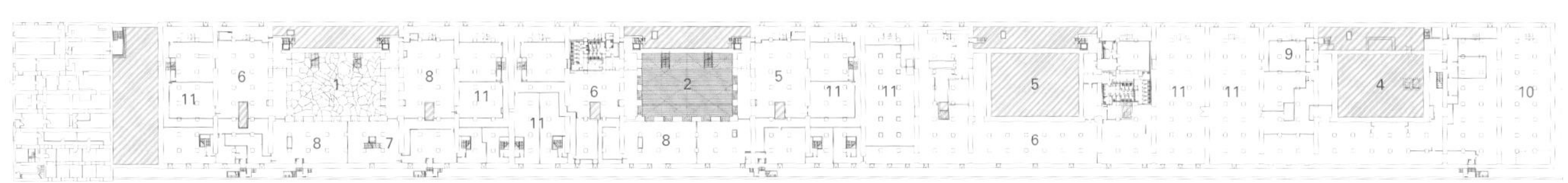

1. 庭院A 2. 庭院B 3. 庭院C 4. 庭院D 5. 设备间 6. 商店 7. 餐厅 8. 咖啡厅/酒吧 9. 文化馆/洛伊斯酒店 10. 市场 11. 贮存室
1. courtyard A 2. courtyard B 3. courtyard C 4. courtyard D 5. services 6. shop 7. restaurant 8. cafe / bar 9. culture / Loisir 10. market 11. stockage
一层下部 lower ground floor

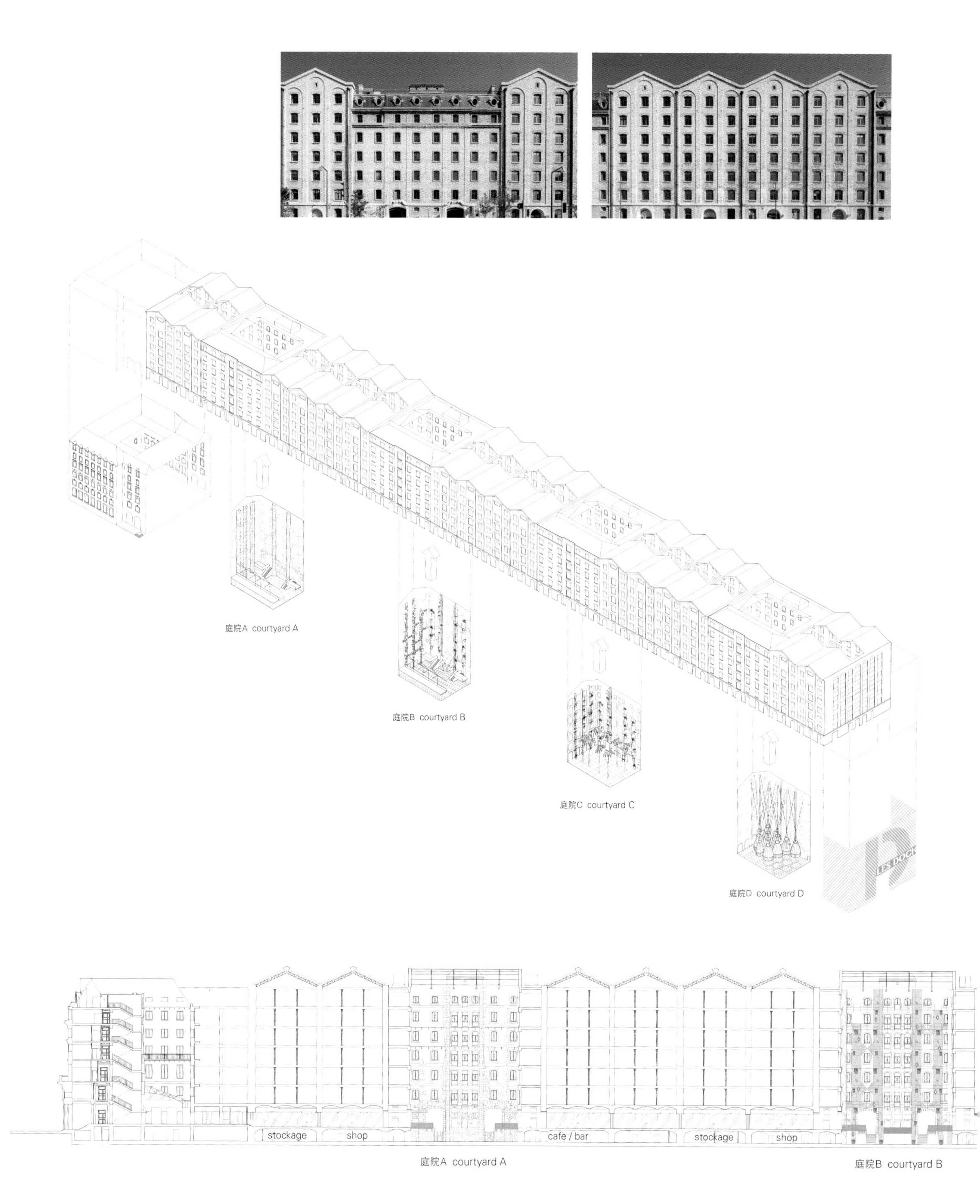

项目名称：Marseilles Docks / 地点：Marseilles, France / 建筑师：Alfonso Femia, Gianluca Peluffo, Simonetta Cenci, Nicola Spinetto_5+1AA / 建筑与景观设计：5+1AA Alfonso Femia Gianluca Peluffo / 项目主管：Nicola Spinetto / 项目负责人：Sara Traverso / 设计团队：Alfonso Femia, Gianluca Peluffo, Simonetta Cenci, Sara Gottardo, Nicola Spinetto, Sara Traverso, Fanélie Pardon, Valeria Parodi, Sara Massa, Luca Bonsignorio, Carola Picasso, Lorenza Barabino / 合作者：Alessandro Bellus, Francesco Busto, Suzanne Jubert, Etienne Bourdais, Gabriele Filippi, M. Cristina Giordani, Valentina Grimaldi, Roberto Mancini, Aude Rasson, Giulia Tubelli, Roxana Calugar

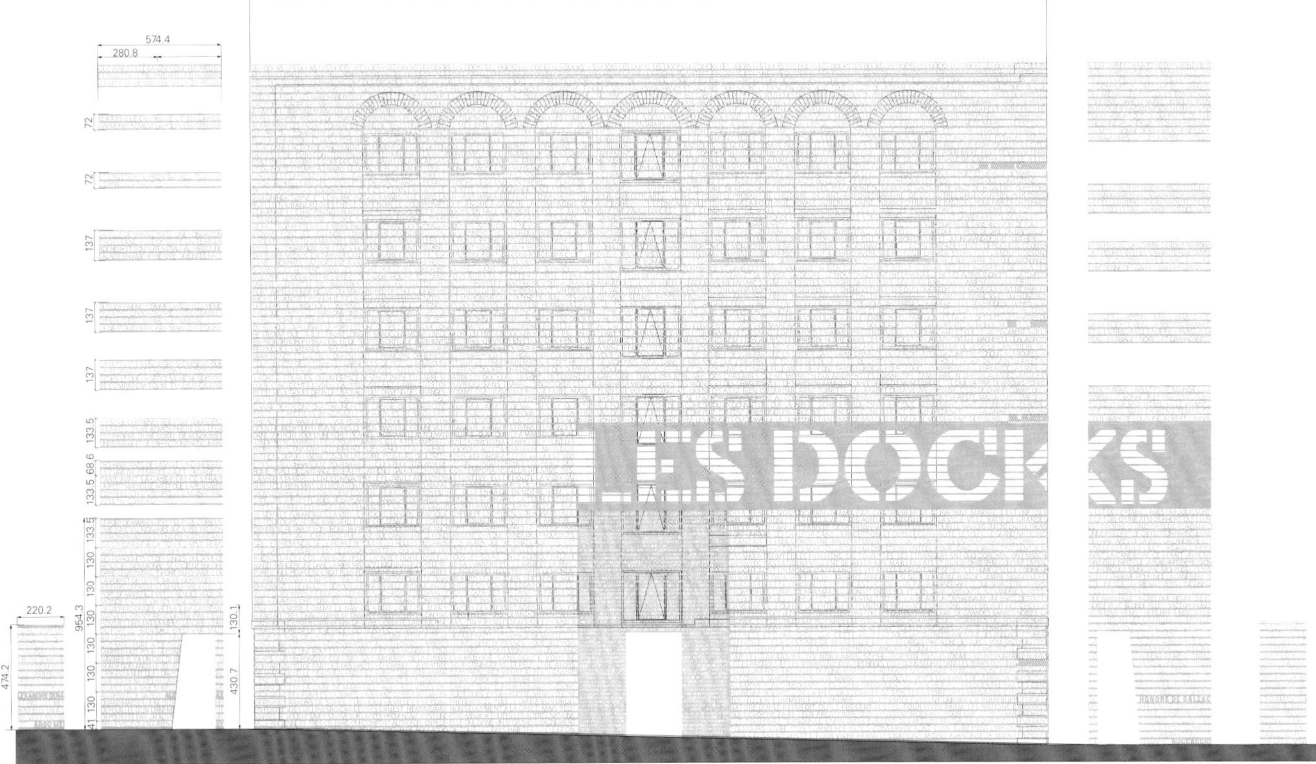

北立面详图 north elevation detail

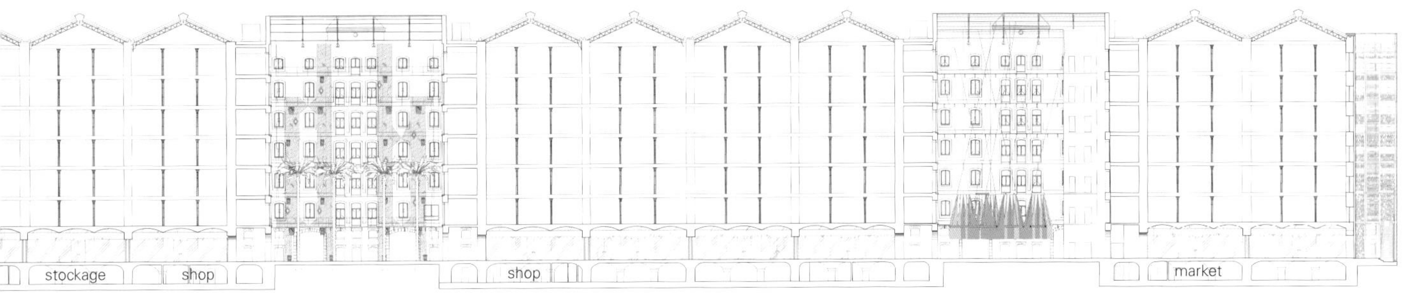

–A' 剖面图 section A-A'　　庭院C courtyard C　　庭院D courtyard D

结构工程师：Secmo / 设备工程师：G2I Garcia / 经济控制办公室：R2M / 控制办公室：Bureau Veritas / 指定承包部门：Constructa Urban Systems / 承包商：Vinci-Dumez Méditerranée et Girard / 瓷砖艺术家：Danilo Trogu, la casa dell'arte – Albissola / 北立面：5+1AA & Tapiro design / 客户：JPMorgan Asset Management, Constructa Urban Systems (Appointed contracting Authority) / 造价：EUR 22,500,000 / 用地面积：23,000m² / 有效楼层面积：21,000m² / 设计时间：2009—2010 / 施工时间：2013—2015.10
摄影师：©Luc Boegly (courtesy of the architect)

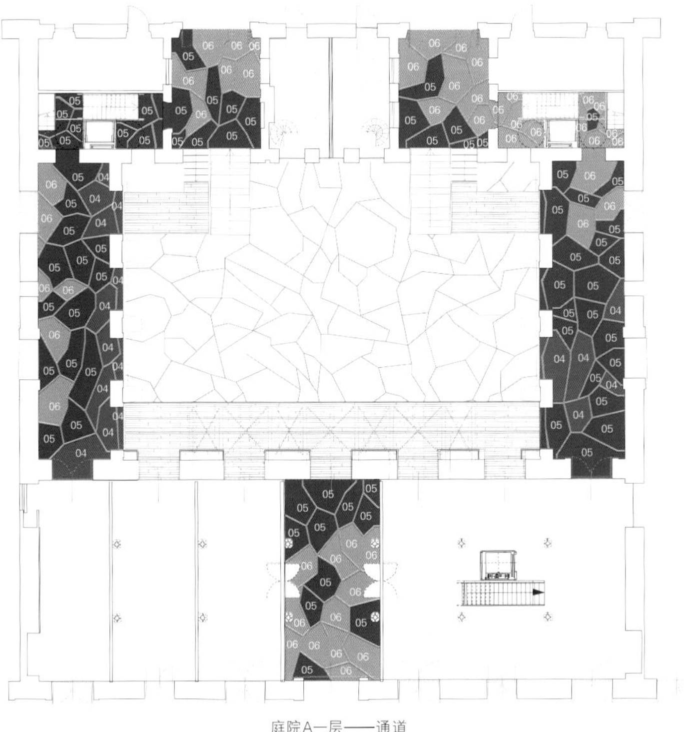

庭院A一层——通道
courtyard A ground floor _ walkways

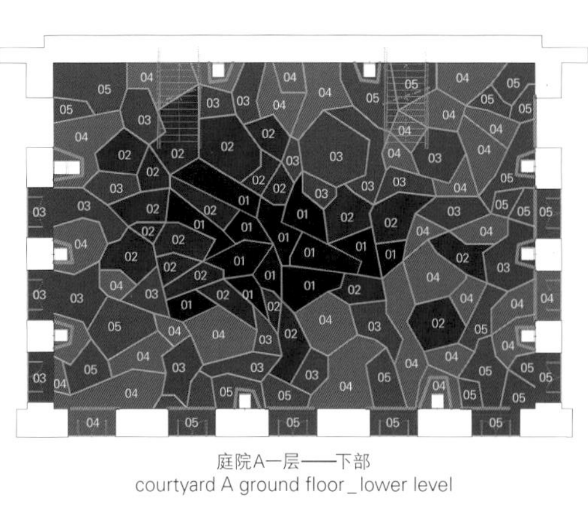

庭院A一层——下部
courtyard A ground floor _ lower level

color code

- B12
- B8
- B4
- B3
- STRONG BLUE
- B1
- ABSOLUTE WHITE

Ceramic Mix 01
- B4=15%
- B8=35%
- B12=50%

Ceramic Mix 02
- B12=15%
- B4=35%
- B8=50%

Ceramic Mix 03
- B8=15%
- B3=35%
- B4=50%

Ceramic Mix 04
- B4=15%
- strong blue=35%
- B3=50%

Ceramic Mix 05
- B3=15%
- B1=35%
- strong blue=50%

Ceramic Mix 06
- strong blue=15%
- B13=35%
- B1=50%

- B1=15%
- absolute white=35%
- B13=50%

Ceramic Mix 08
- B13=50%
- absolute white=50%

Ceramic Mix 08
- absolute white=100%

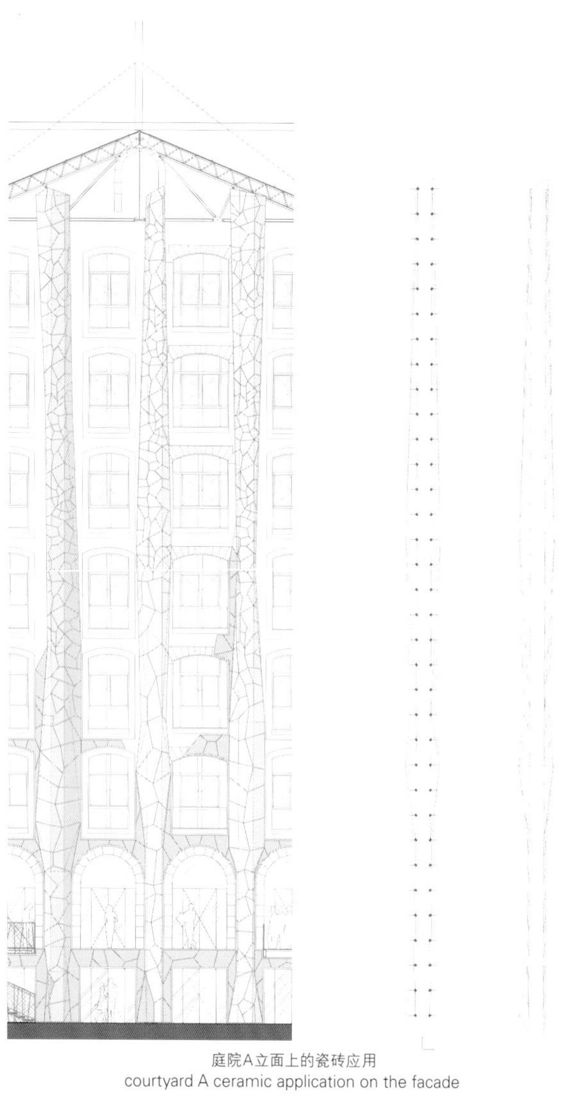

庭院A立面上的瓷砖应用
courtyard A ceramic application on the facade

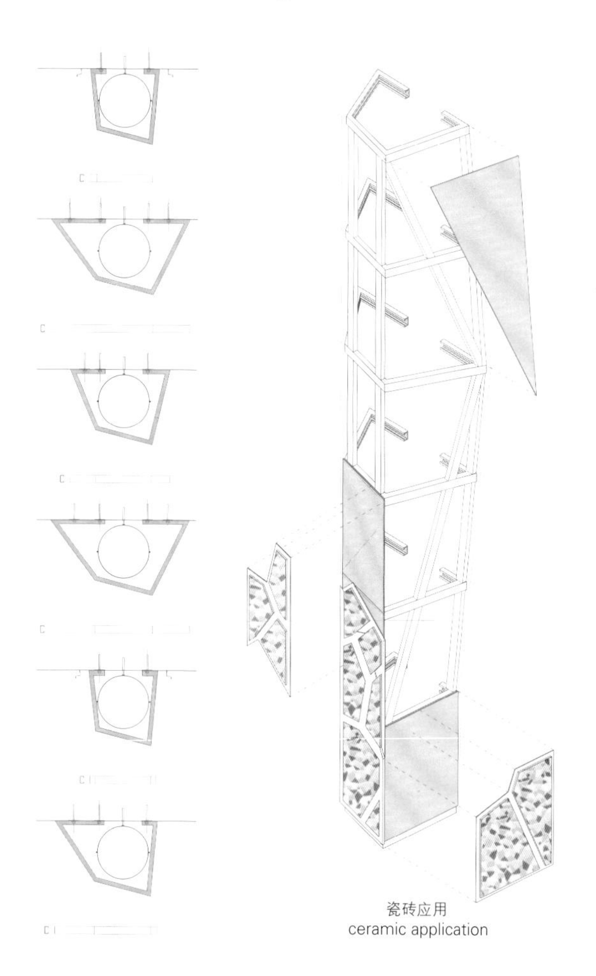

瓷砖应用
ceramic application

ated place gains continuity to the sea. The project redefines the four courtyards destined to be commercial, artistic and cultural that, from inside, extend towards the two exterior places and the two adjacent streets. They create four breaks inside a transverse route.

The Harbor, the Village, and the Market, the three themes that punctuate the new spaces of this symbolic building of Marseilles is well represented by surprising use of materials, unexpected sceneries, and adapted plants.

The materials and spatial elements used are symbolic and monumental, as with the building itself. They are reinterpreted into an intimate sentiment and wonder. Also the unique ambiance each of the four courtyards create, is presented by the change and mingling of different everyday lives. Furthermore, a new relationship with time is set between the space and its users, based on exact calculations of the seasonal calendar.

Time and nature are reflected through the ceramic materials of the four courtyards. Light is controlled in accordance with the passing of the sun, letting different colors and scents to flow into the emptied space. A dialogue is immediately made between nature and artifice, the sensed and the existent, emotion and reason. Meanwhile the everyday life with all its hustle and bustle is filtered and projected by screens and drapes arrayed. These and various different elements create a space that is unable to be indifferent towards the city and its people.

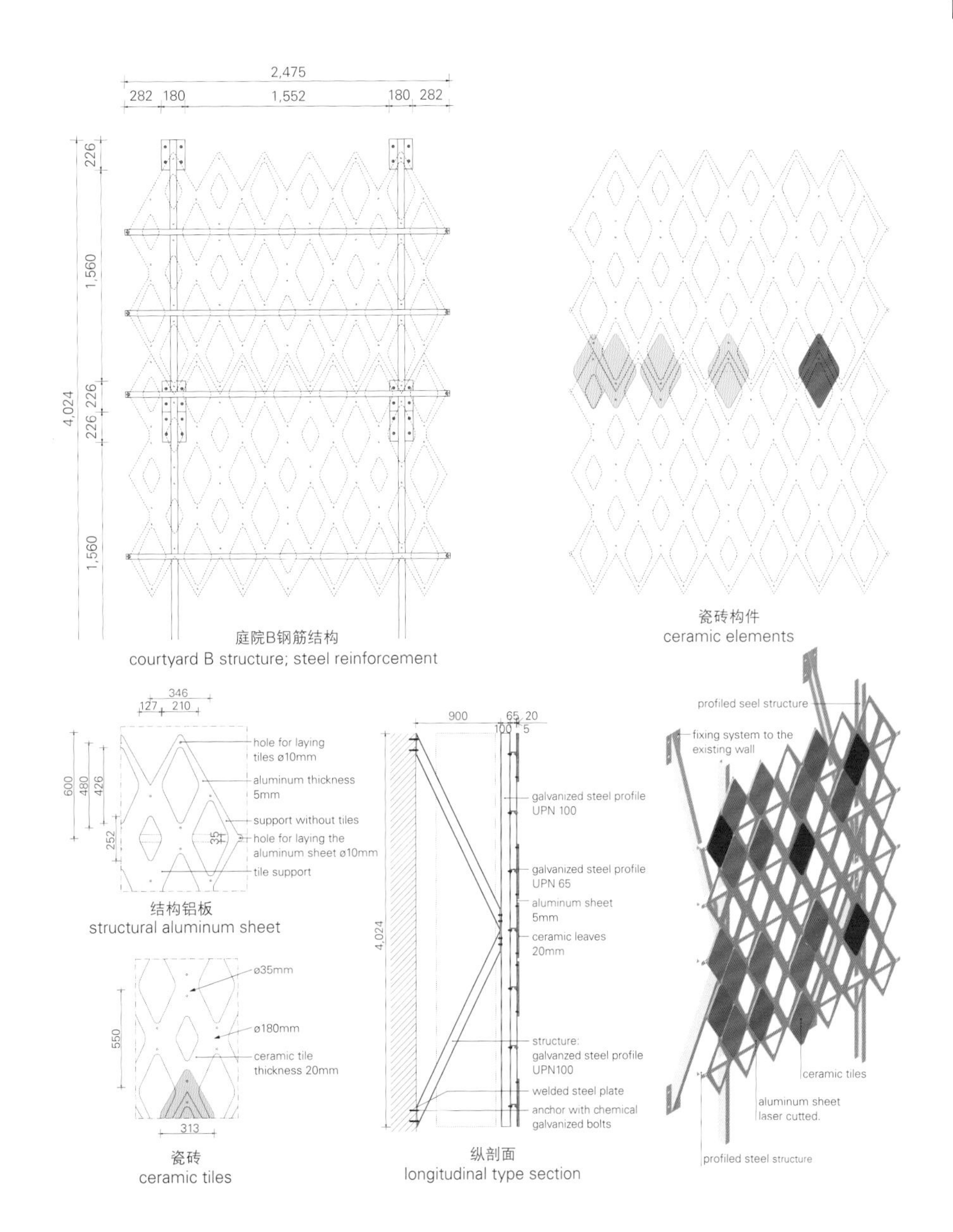

庭院B钢筋结构
courtyard B structure; steel reinforcement

瓷砖构件
ceramic elements

结构铝板
structural aluminum sheet

瓷砖
ceramic tiles

纵剖面
longitudinal type section

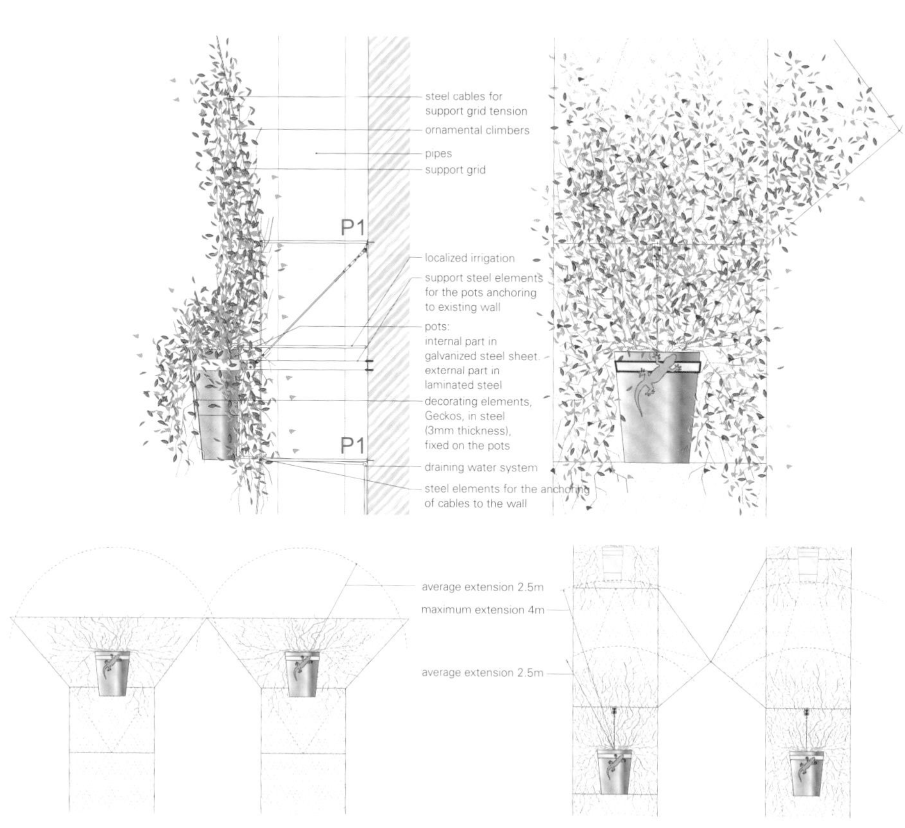

庭院C装饰性攀缘植物的最大攀爬效果
courtyard C maximum extension of ornamental climbers plants

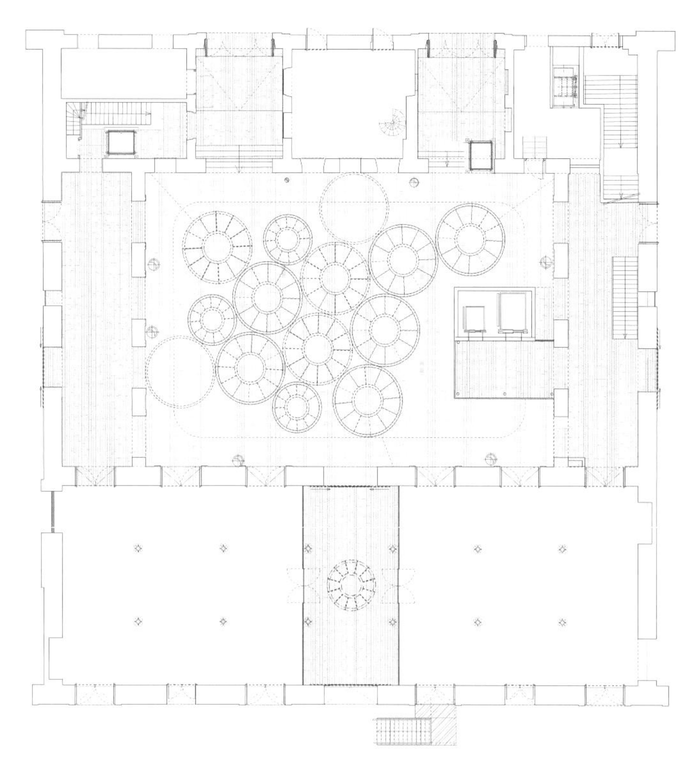

庭院D灯具与圆形结构
courtyard D lights and circus structures

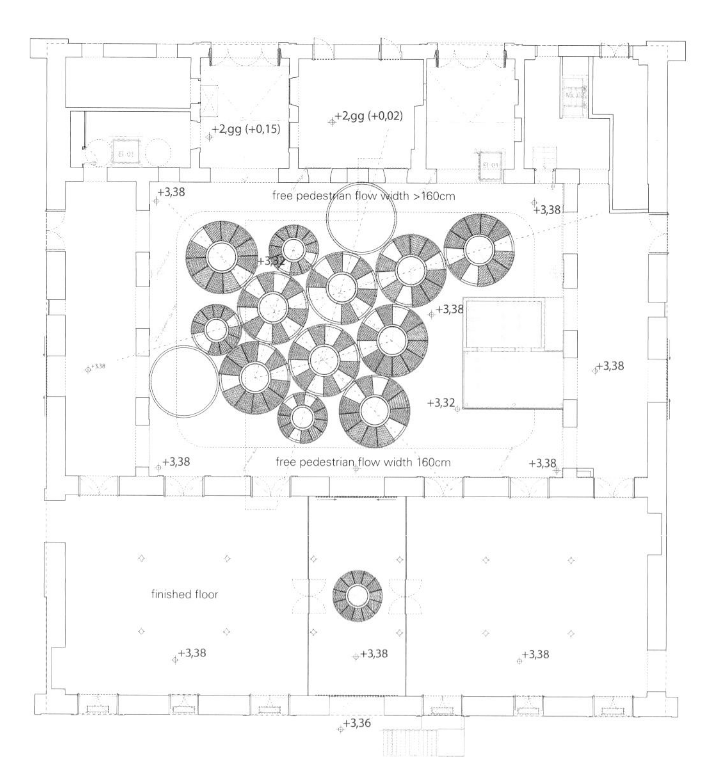

庭院D金属板分布
courtyard D sheet metal distribution

Le fournil

街头艺术博物馆

Studio ARCHATTACKA

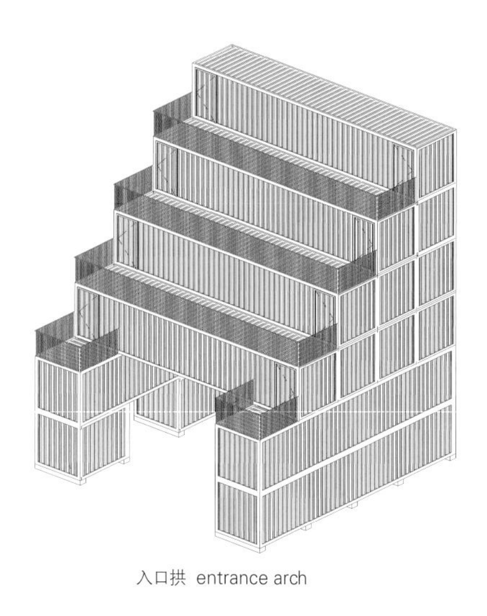

入口拱 entrance arch

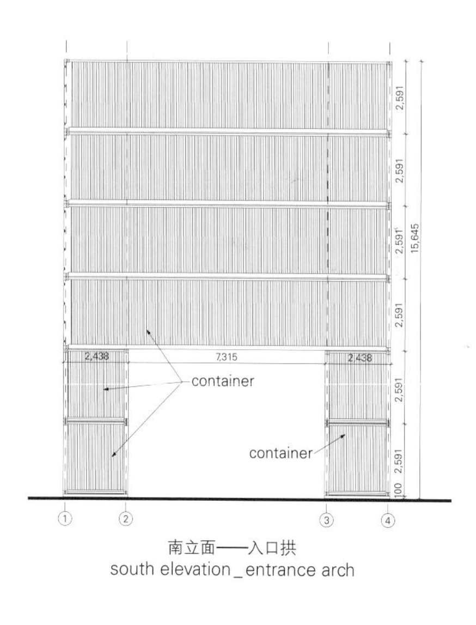

南立面——入口拱
south elevation _ entrance arch

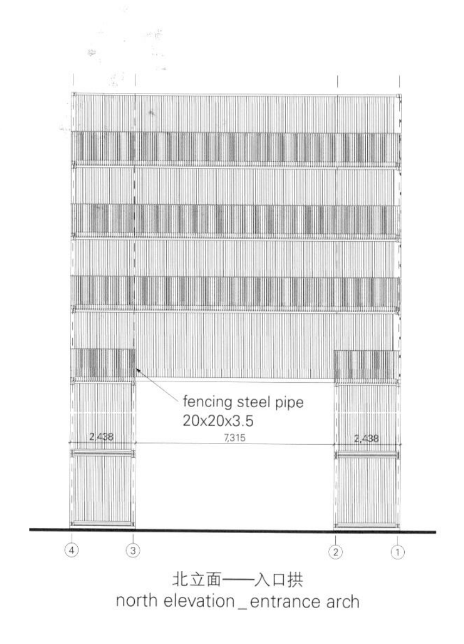

北立面——入口拱
north elevation _ entrance arch

1. 演讲大厅
2. 儿童游乐场
3. 街头球场
4. 轮滑坡道
5. 乒乓球桌
6. 礼品店
7. 公共厕所
8. 露天剧场
9. 金库
10. 景点
11. 原有的飞机乘客舷梯
12. 展厅
13. 酒吧
14. F座入口
15. F座
16. 入口大厅
17. 市场
18. 展厅（主管道）
19. 停车场

1. lecture hall
2. children playground
3. street ball playground
4. skate ramp
5. pingpong table
6. gift shop
7. water closets
8. amphitheater
9. cashbox
10. scene
11. old aircraft passenger loader
12. exhibition halls
13. bars
14. building F entrance
15. building F
16. entrance hall
17. market
18. exhibition hall (main tube)
19. parking

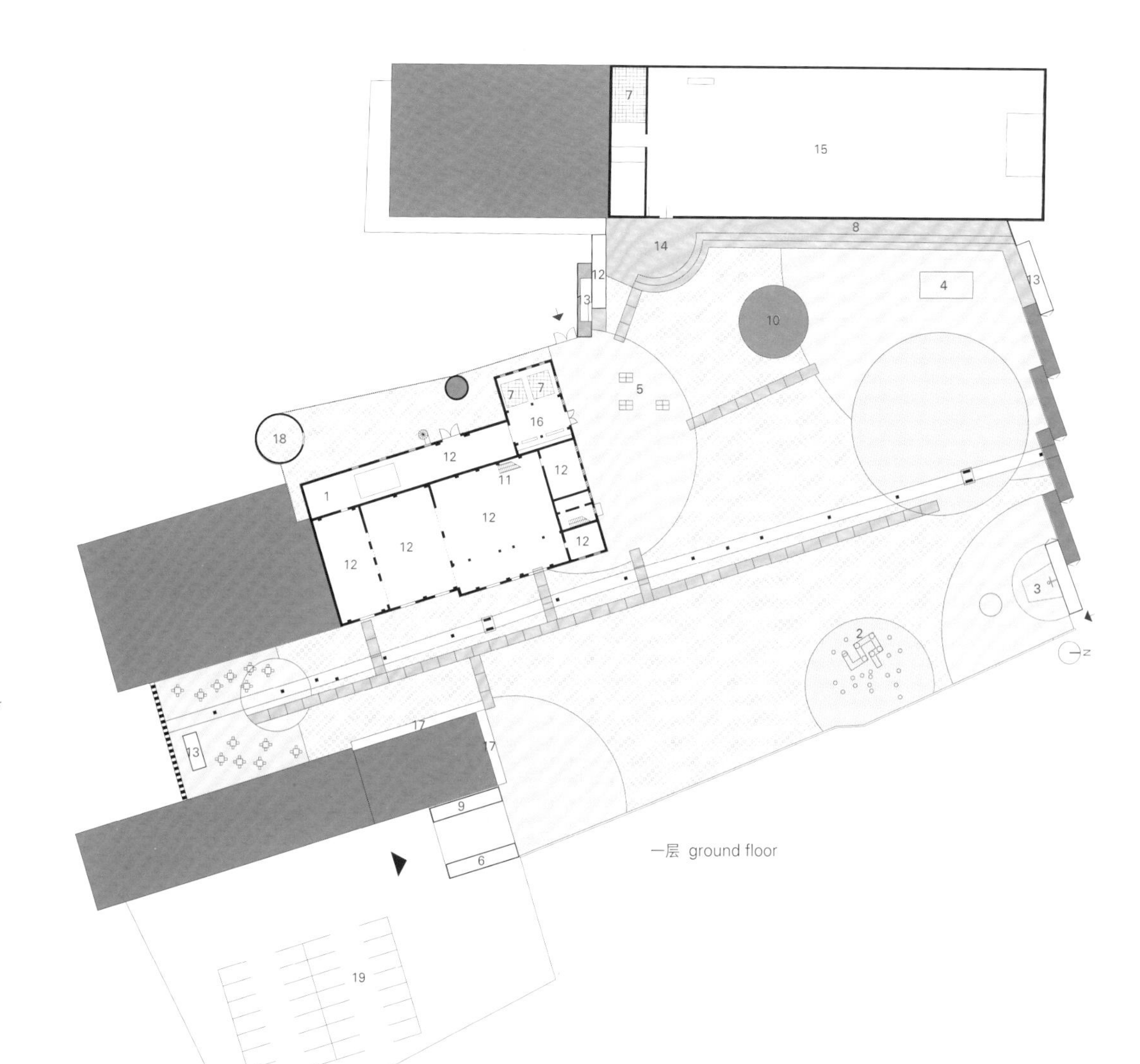

一层 ground floor

НАЗА
НЕЛЬ

俄罗斯圣彼得堡的HPL塑料厂工业区改建是一个大项目，而这个大项目的第一步便是街头艺术博物馆的设计与建造。自三年前建成以来，这个有着70年历史的建筑已经举办了四次大规模的展览，来自世界各地的艺术家们纷纷来到此处进行参观与交流。同时，它也为许多小型节日提供庆祝场地。现在，这个博物馆已经成为该地区一个新的文化中心。

乍一看，这个工厂与艺术毫不相干，但实际上，它一直以来都与艺术密切相关。年轻的艺术家们为了寻找廉价的地方建立工作室，聚集在这个废弃的工厂里，并经常把它用于他们所创作的艺术作品中。事实上，在2015年年底，法国艺术家伯兰特·戈斯林曾举办了一场名为"艺术创作/生产"的展览，就向人们展示了这个工厂与艺术之间的独特联系。该展览主要展示了艺术与其平台之间的概念和流程。

在改建工厂与其旧址时，需要考虑许多因素。该项目必须能够为展览提供灵活而临时的展览空间；而在夏天，这儿又需要用来举办各种节日庆祝活动；与此同时，还要保证原有建筑的空间原貌。但这些限制条件最终使设计成为有意义、有创造性的过程，艺术元素大量地融入了建筑设计中。此外，紧张的预算要求采取能节约成本的对策，因此该项目采用了预制和回收利用的方法。回收利用的材料以及所有用于施工的材料，在应用时都考虑到有朝一日能够再次被回收利用。

该项目的设计目的不是设计一个"建筑艺术品"，而是将这个工业空间及其历史打造成一个艺术品。一些传统的机构也许更倾向于永久性建筑，其坚固的形式更利于展现建筑景观。然而，在这里，建筑师用一个灵活而朴素的空间来突显艺术作品。这样的设计使得这些简单、清晰但千变万化的空间可以用来举办各种展览。有趣的是，只有光滑平整的草坪才会体现出建筑师的存在。在这个空间里，每一个元素都被巧妙地设计成独一无二的独立个体，客人无处不在，而客人本身就是艺术。

Street Art Museum

The Street Art Museum project is the initial part of a bigger project that transforms the industrial district of the HPL plastic factory in St. Petersburg, Russia. After its completion 3 years ago, the 70 year-old building has become a platform for four large-scale exhibitions in which a large number of artists from all over the world participated. Also, as a setting for many small festivals, the museum is now a new cultural center for the area.

What may at a glimpse look unrelated, the factory has actually been in close relation with art. Young artists looking for a cheap place to set up studio, gathered to this abandoned factory and frequently used it in their creative works of art. In fact, French artist Bertrand Gosselin held an exhibition in late 2015, which presented this unique connection. His exhibition titled "Art Production/ Production" characterized the concepts and processes between art and its platform.

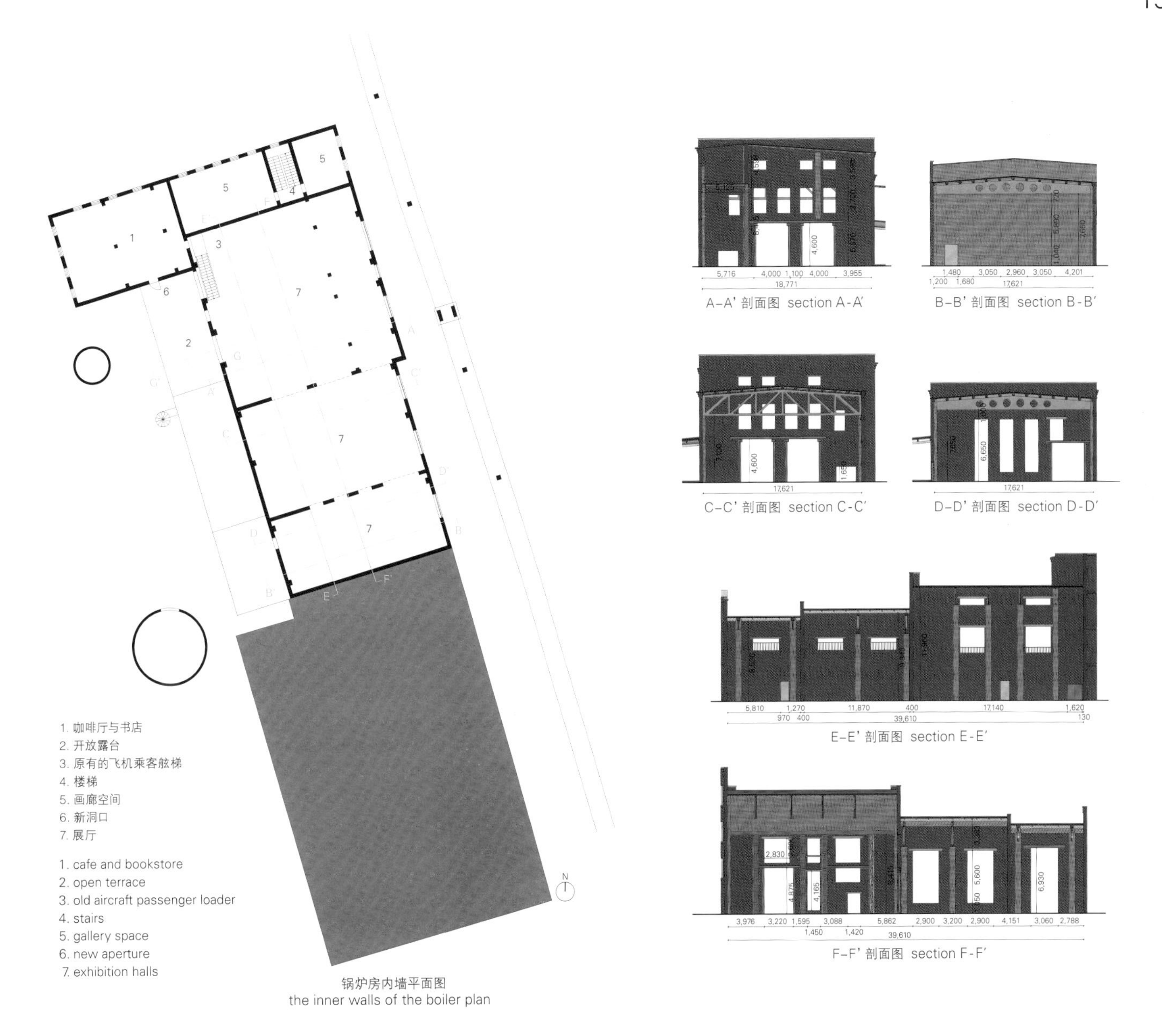

1. 咖啡厅与书店
2. 开放露台
3. 原有的飞机乘客舷梯
4. 楼梯
5. 画廊空间
6. 新洞口
7. 展厅

1. cafe and bookstore
2. open terrace
3. old aircraft passenger loader
4. stairs
5. gallery space
6. new aperture
7. exhibition halls

锅炉房内墙平面图
the inner walls of the boiler plan

A–A' 剖面图 section A-A'

B–B' 剖面图 section B-B'

C–C' 剖面图 section C-C'

D–D' 剖面图 section D-D'

E–E' 剖面图 section E-E'

F–F' 剖面图 section F-F'

项目名称：Street Art Museum / 地点：Shosse Revolyutsii 84, St. Petersburg, Russia
建筑师：Studio ARCHATTACKA / 建筑师负责人：Alexander Berzing, Andrey Voronov
承包商：Laminated Plastics plant "SLOPLAST" / 客户：Laminated Plastics plant "SLOPLAST"
用途：museum, cultural center, gallery, library
用地面积：12,260m² / 建筑面积：220m² / 结构：3,680m²
材料：old sea containers, steel frame / 造价：steel, wood, concrete, bricks, stones, hpl -plastic
设计时间：about 300,000~350,000 EUR (for three years)
竣工时间：2013—2016 / Completion: 2016
摄影师：©Alexander Berzing (courtesy of the architect) - p.150, p.157[bottom], p.158[top], p.159[bottom], p.161[bottom-left, bottom-right]
©Alexander Sigaev (courtesy of the architect) - p.152~153, p.154, p.156, p.157[top], p.158[bottom], p.159[middle], p.161[top, middle]
©Denis Batishev (courtesy of the architect) - p.151
©StreetArtMuseum (courtesy of the architect) - p.155, p.159[top], p.160

1. 演讲大厅
2. 带有储存家具的围合构件
3. 展厅
4. 接待处
5. 公共厕所

1. lecture hall
2. exposure unit with storage furniture
3. exhibition hall
4. reception
5. water closets

CAI
Константин
Новиков
БИБЛИОТЕКА

CAI

CAI
CAI
CAI
CAI
CAI
CAI

Many factors were needed to be considered in renovating the factory and its site. The project had to be able to accommodate flexible and temporary exhibition spaces that could also be used for the summer festivities, while keeping intact the existing spatial qualities. However, these restrictions only developed the design into a meaningful creative process, actively inviting art in architecture. Furthermore, a tight budget called for cost-efficiency measures which brought the project to incorporate pre-cast and recycling methods. Salvaged materials, or anything that could be used for construction, were applied in way capable to be reclaimed again someday. Rather than designing an "architectural oeuvre", the project focuses on transforming the industrial space and its history into a massive land art object. Traditional institutions would have preferred the rigid form of a permanent structure to stand out and represent the landscape. However, here a flexible unostentatious space is created to highlight the pieces of art. In doing so, the ever-changing spaces, although simple and clear, will be able to host whatever exhibition. Interestingly enough, only the smooth lawn shows the presence of an architect. It is art, the guest, that prevails in the space, through every element subtly and holistically designed as unique independent objects of a composition.

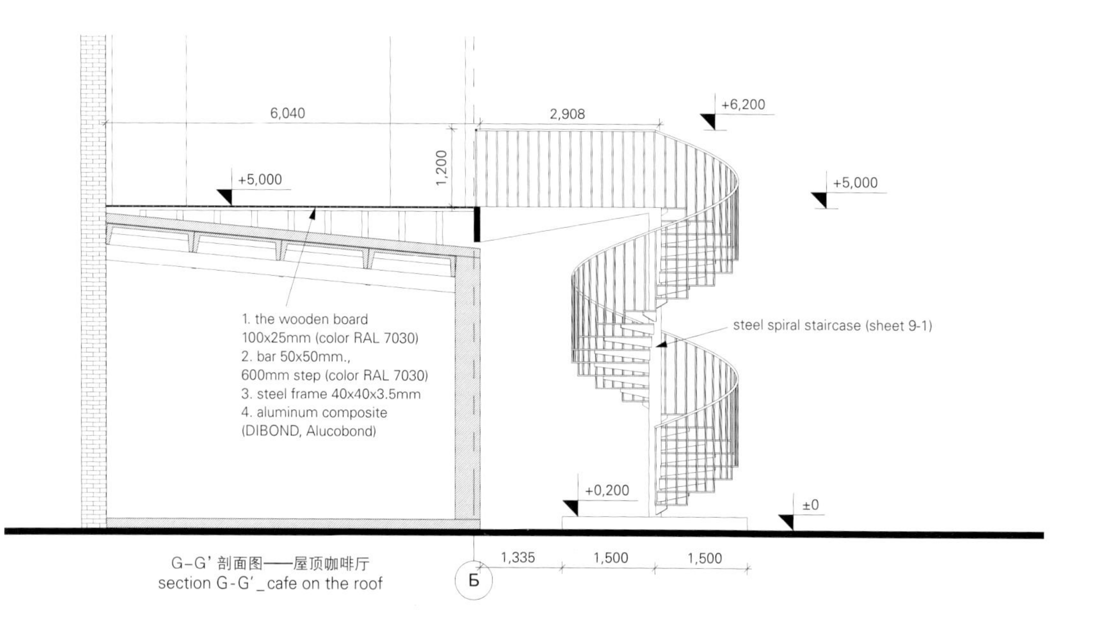

G-G' 剖面图——屋顶咖啡厅
section G-G'_cafe on the roof

"Я ПОТЕРЯЛ СВОЮ ПАМЯТЬ"

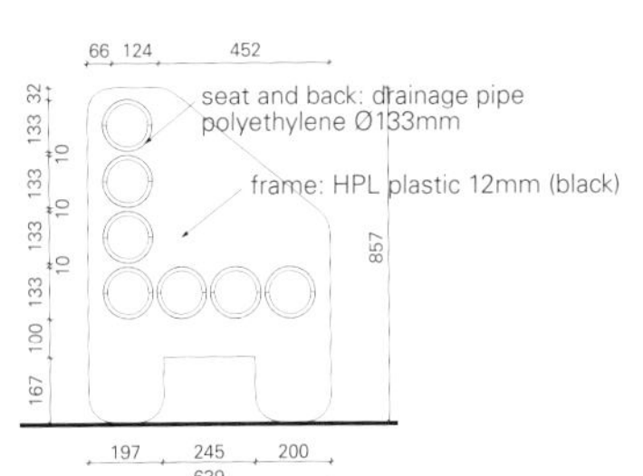
66 124 452
seat and back: drainage pipe
polyethylene Ø133mm
frame: HPL plastic 12mm (black)
857
197 245 200
639

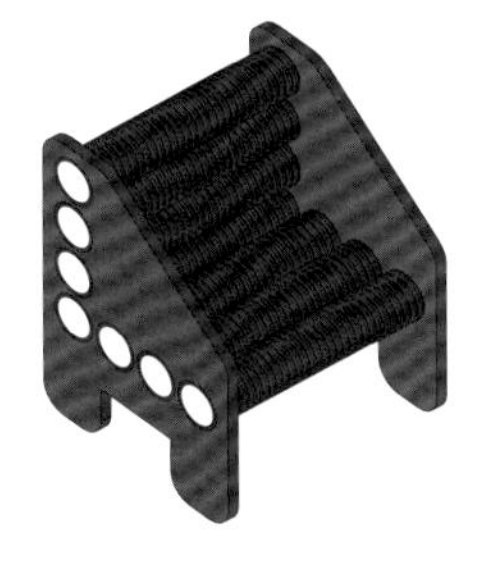

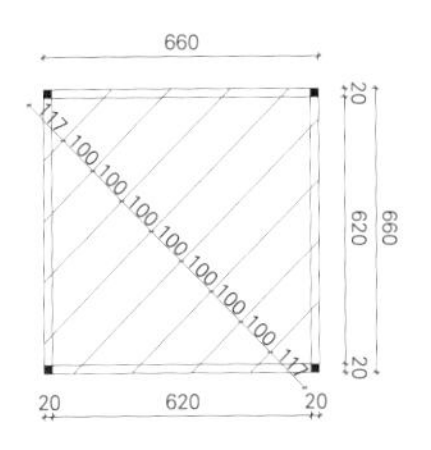
660
620
20 620 20

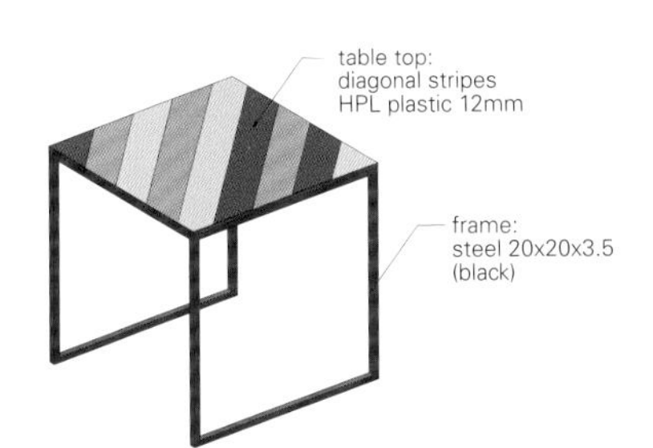
table top:
diagonal stripes
HPL plastic 12mm
frame:
steel 20x20x3.5
(black)

伯明翰新街火车站

AZPML

伯明翰新街火车站是一个重要的交通枢纽，也是城市公共领域的一个关键点。新街火车站占据着城市的重要位置，每天的交通吞吐量十分巨大，给大量要去英格兰中西部地区的旅客留下对伯明翰市的第一印象。

伯明翰新街火车站旨在成为一座标志性建筑，不仅要在人们心中留下深刻印象，而且要向公众传达此建筑的功能和其处在伯明翰市中心的地理位置特点。为了实现这一目标，新街火车站设计方案要表达的是铁路主题的动态特性。动态的几何形体和由移动产生的感知变形扭曲成为本项目建筑表现的灵感。

火车站铁轨那分叉、起伏、顺滑的形态被转移并具体体现在该项目的几何建筑中，使城市变得华美绚丽，同时表达了其作为交通枢纽的历史性特征。各种交通系统，像著名的运河和罗马道路等，都在此处汇聚，彼此交织、覆盖、叠加。此设计的目的在于，通过特别明确表现所精心挑选的几处火车站附近城市景观，触发人们对火车站周围城市环境的全新感知。

新火车站设计无论是在建筑的外观上还是在其重组上，其设计方

法都旨在重新建立形式与设计理念之间的一致性。无论从组织结构还是从视觉方面来说，旧火车站所要达到的效果与新火车站设计所要达到的效果是不可同日而语的。实际上，建筑的外立面与内部构造是不相关的，恰恰相反，其外立面设计与建筑外部空间是息息相关的，从而加强了人们对伯明翰市中心生活的感知,而不是力图揭示其内部构造。对于处理当代文化的复杂性，该设计方法是十分必要的，它把现代主义者所提倡的透明性演变为一种更务实、更具策略性的方法。

该设计将外部的防雨屏转换成可反光的不锈钢曲面。如此这般设计，是有目的地映射周围城市环境，映射曾经灰暗如今明亮的伯明翰市天空，映射熙熙攘攘的人群和不断进出车站的火车，映射日出的光芒和日落的余晖，映射此处其他富有活力的东西。为了突出火车站四个主要入口，入口处的建筑外立面上安装了形如眼睛的大型多媒体电子屏幕，与建筑外立面融为一体。

该建筑外部的反射面使整栋建筑外表看起来连贯一致，但因反射面位于建筑的不同侧面，其反射的场景也是有所不同的。

Birmingham New Street Station

Birmingham New Street Station is an important transport hub and a key aspect of the city's public realm. Occupying an important position in the city and handling a large amount of traffic, it provides the first impression of Birmingham to a large influx of visitors to the Midlands.

The proposal for Birmingham New Street Station seeks to produce an iconic architecture that, beyond creating an impression, will be able to communicate the function of the building and the character of its location at the very centre of Birmingham City, to the public. To achieve this it is proposed to provide expression to the dynamic nature of the railway theme. The geometries of motion and the distortion of perception produced by movement have been the inspiration for the architectural expression of the project.

The bifurcating, undulating, smooth forms of the track field have been transferred and embedded into the geometry of the building to ornate the city and to convey its historical character as a transportation hub, where various traffic systems – such as the famous canals, the roman roads etc. converge and overlay. The design aims to trigger a new perception of the urban settings around the station, by specifi-

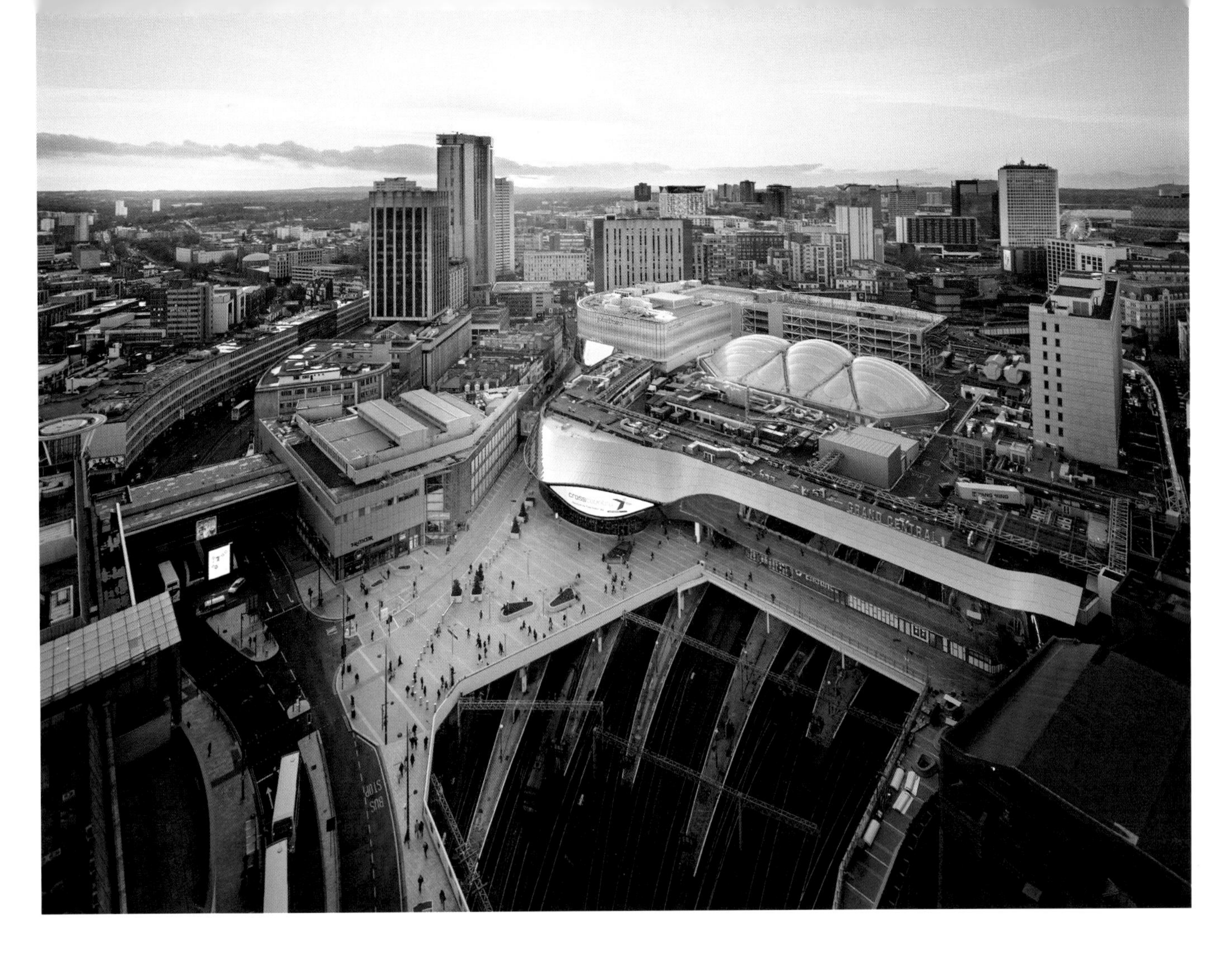

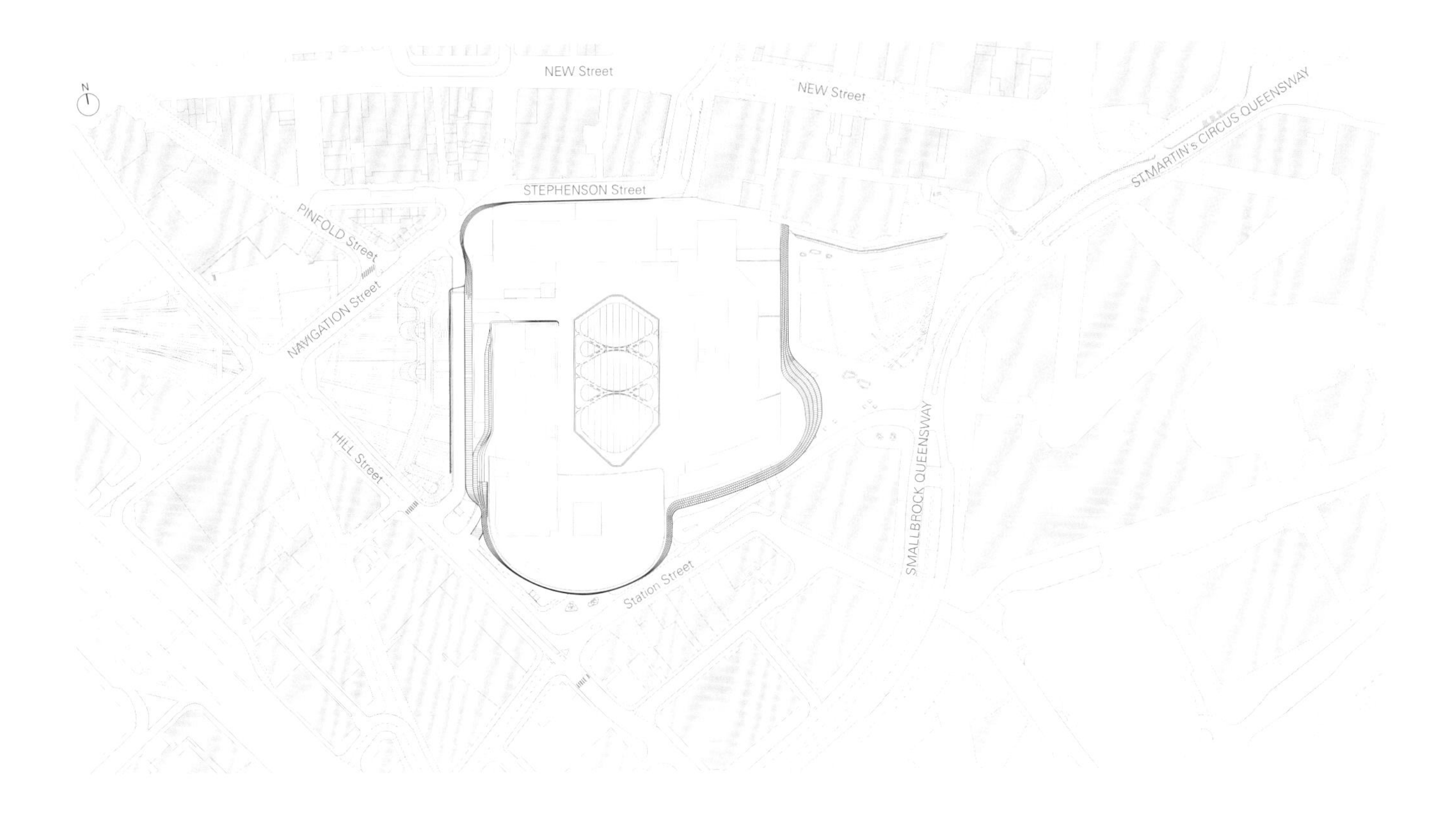

项目名称：Birmingham New Street Station / 地点：Birmingham, UK / 建筑师：AZPML / 执行建筑师：Atkins / 合伙人负责人：Alejandro Zaera-Polo
项目团队：Manuel Eijo, Guillermo Fernandez-Abascal, Charles Valla, Christof Trenner, Tommaso Franzolini, Lola Fernandez, Sukyeong Kim, Carmen Sagredo, Takeru Sato, Penny Sperbund, Niklavs Paegle, Tobias Jewson, Mio Sato, Manuel Távora / 机电顾问：Hoare Lea (MEP), Atkins / 质量监督：Faithful & Gould / 分包商和供应商：NG Bailey, Coleman & Company, Elliott Thomas, Martifer UK, Fireclad, MPB, SAS, Vector Foiltech, Glazzard / 项目经理/总承包商：MACE / 有效楼层面积：91,500m^2
(lower mezzanine: 3,000m^2; platforms: 8,000m^2; concourse: 20,000m^2; upper mezzanine: 4,500m^2; grand central: 17,000m^2; JLP: 24,000m^2; upper retail: 15,000m^2) / 造价：£750 million / 施工时间：2009—2015.9 / 摄影师：©Javier Callejas (courtesy of the architect)

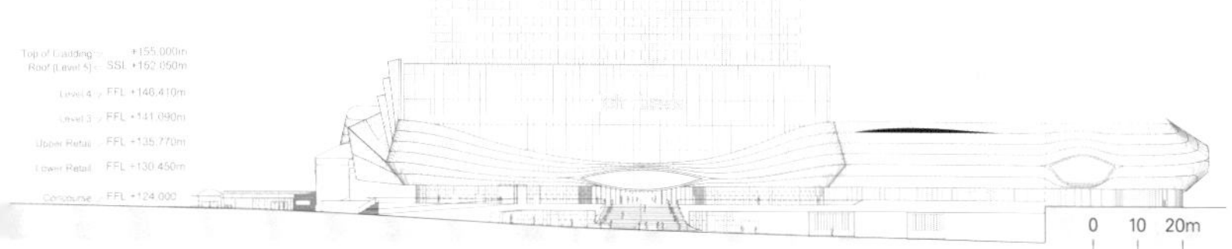

南立面 south elevation

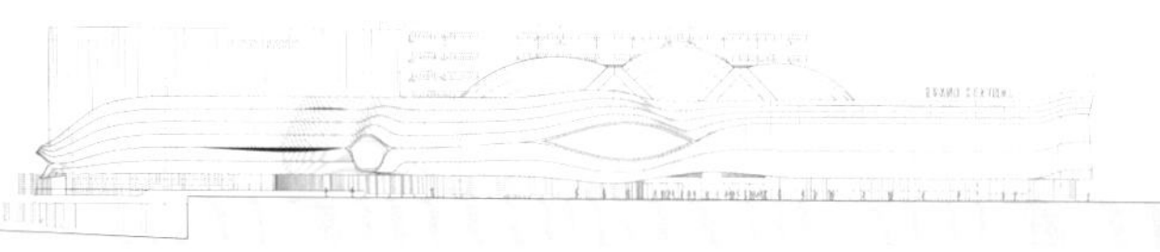

东立面 east elevation

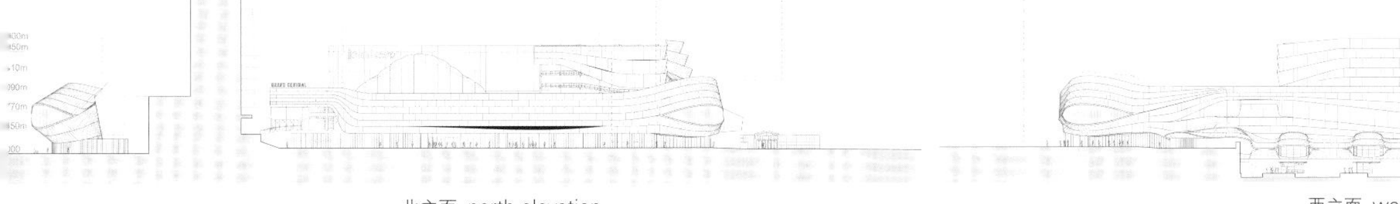

北立面 north elevation

西立面 west elevation

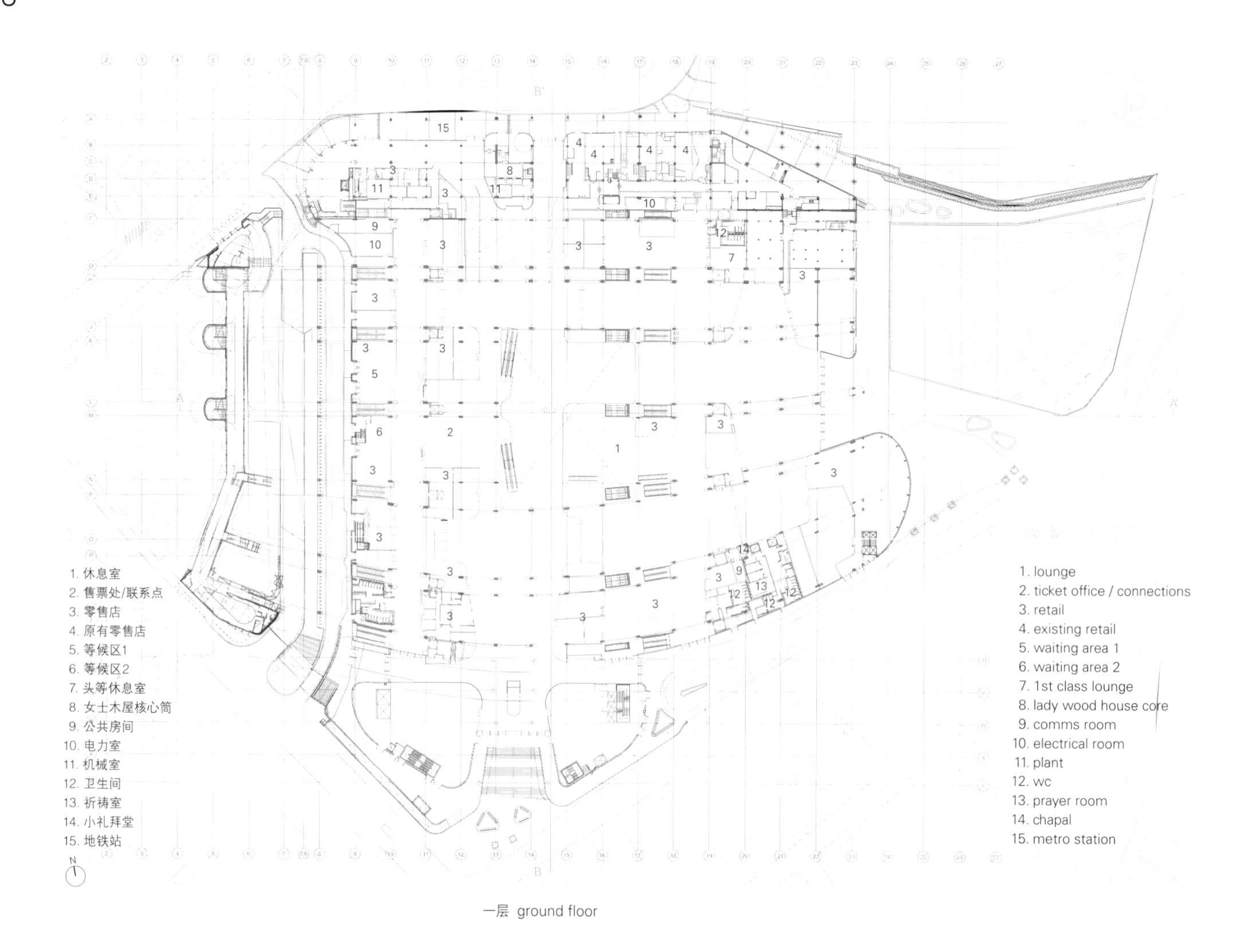

一层 ground floor

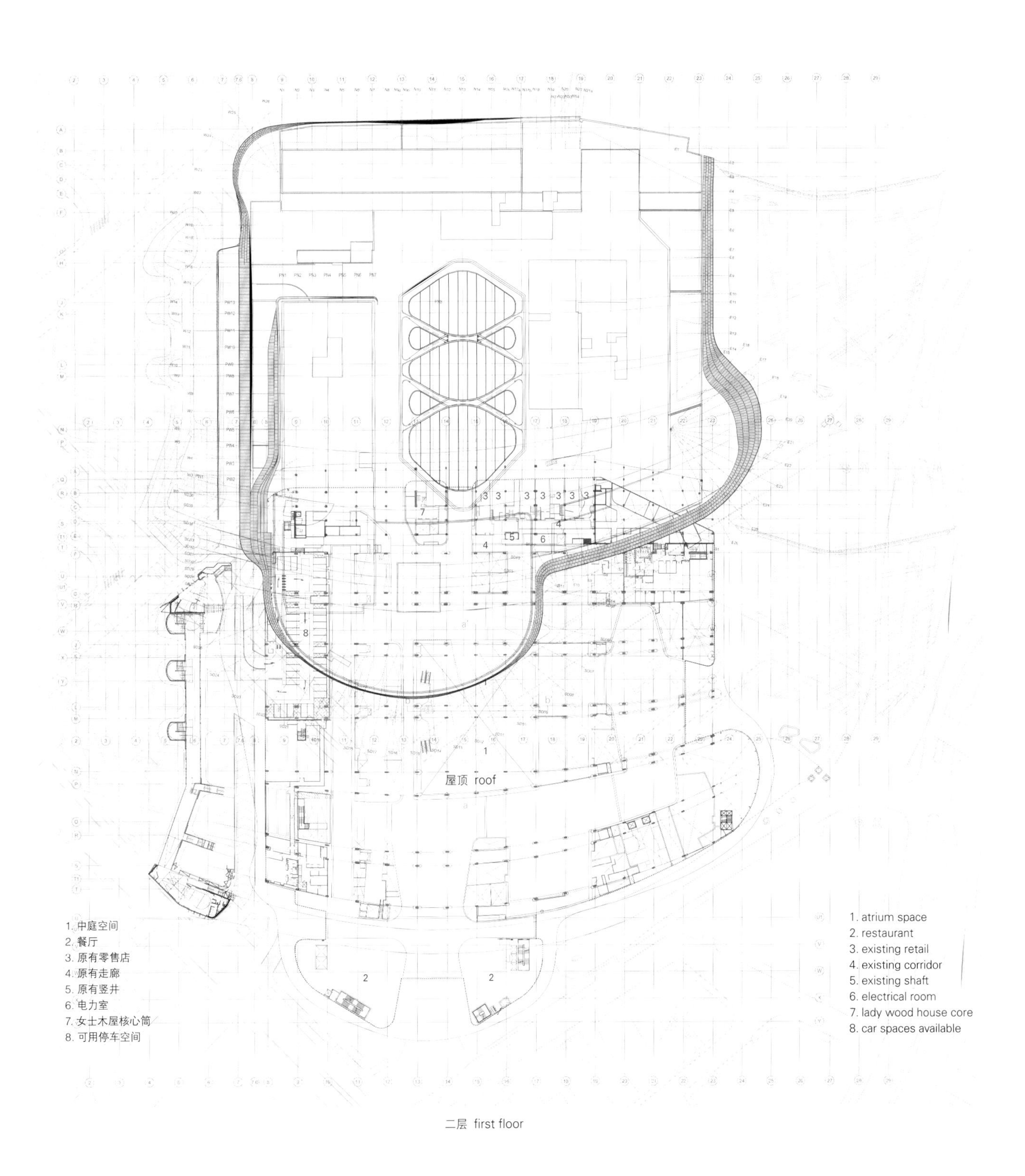
屋顶 roof
1. 中庭空间
2. 餐厅
3. 原有零售店
4. 原有走廊
5. 原有竖井
6. 电力室
7. 女士木屋核心筒
8. 可用停车空间
1. atrium space
2. restaurant
3. existing retail
4. existing corridor
5. existing shaft
6. electrical room
7. lady wood house core
8. car spaces available

二层 first floor

cally reflecting selected areas of the urban landscape around the station.

The design approach aims to re-establish consistency between form and expression in the new station design, both in the cladding and in the re-organisation of the building. The old structure of the building was built for a different performance to the one is now being sought, both in organisational and visual terms. As the cladding cannot be related to the interior of the building for practical reasons, the design of the facade has been related to the exterior space, making the building an instrument to intensify the perception of urban life in Birmingham's inner city, as opposed to try to reveal its inner structure. This approach is a necessary evolution of the modernist dogmas of transparency towards a more pragmatic and strategic approach necessary to address the complexities of contemporary culture.

By turning the external rain screen into a warping, reflective stainless steel surface, Birmingham New Street Station will be designed to produce controlled reflection of the surrounding urban field to reflect the once dark, now bright Birmingham sky, the crowds of passengers, the trains entering and exiting the station, the hues of the sunset and sunrise, and other dynamic regimes present at the site. To highlight the four main access points, large "eye-shaped" media screens have been integrated in the facade.

The field of reflections which constitutes the external envelope of the building, and produces a consistent identity, differentiates depending on the opportunities on each side of the building.

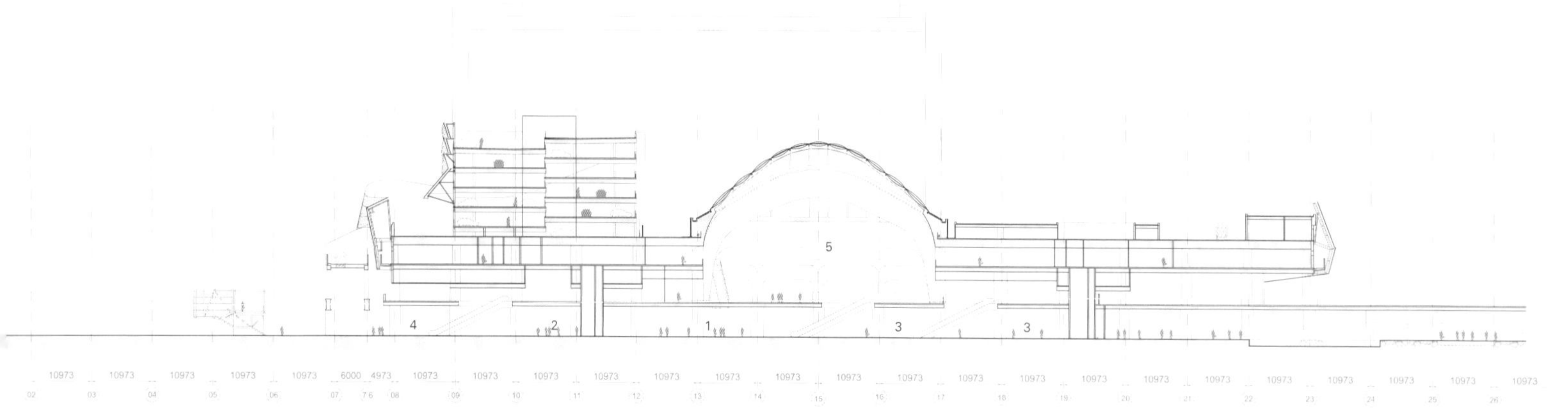

1. 休息室 2. 售票处/联系点 3. 零售店 4. 等候区 5. 中庭空间 1. lounge 2. ticket office / connections 3. retail 4. waiting area 5. atrium space
A-A' 剖面图 section A-A'

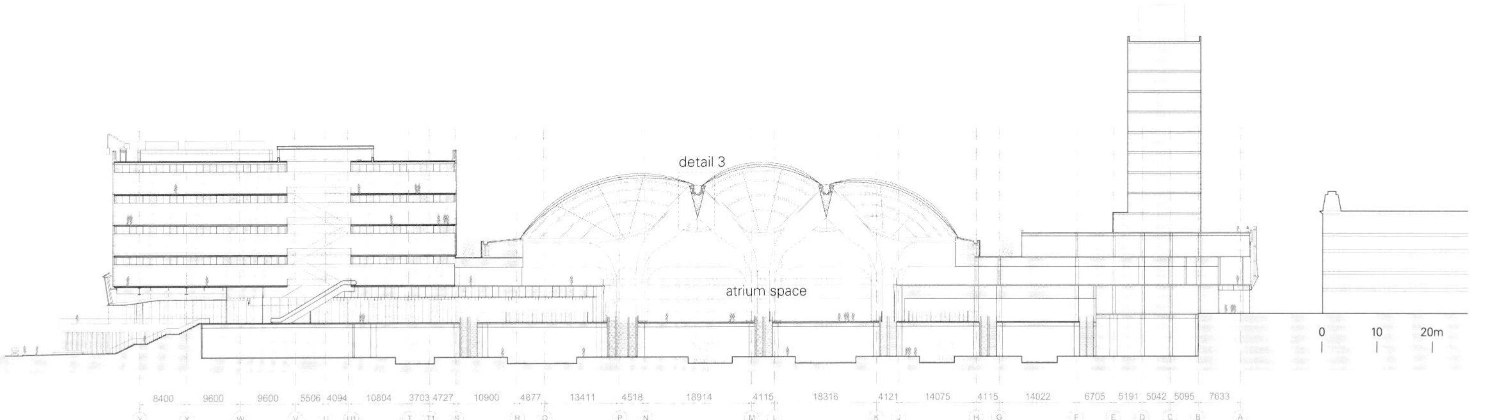

B-B' 剖面图 section B-B'

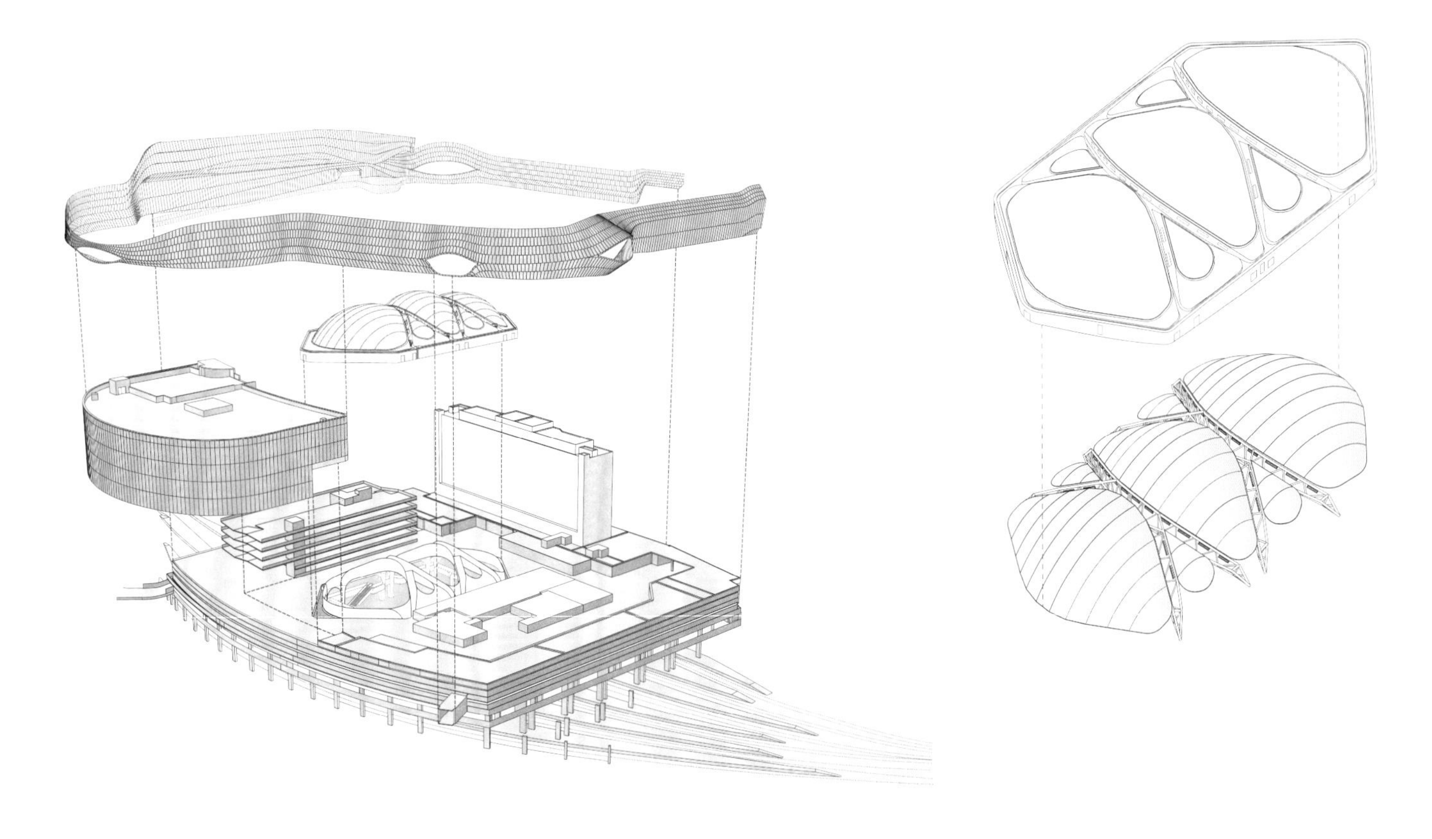

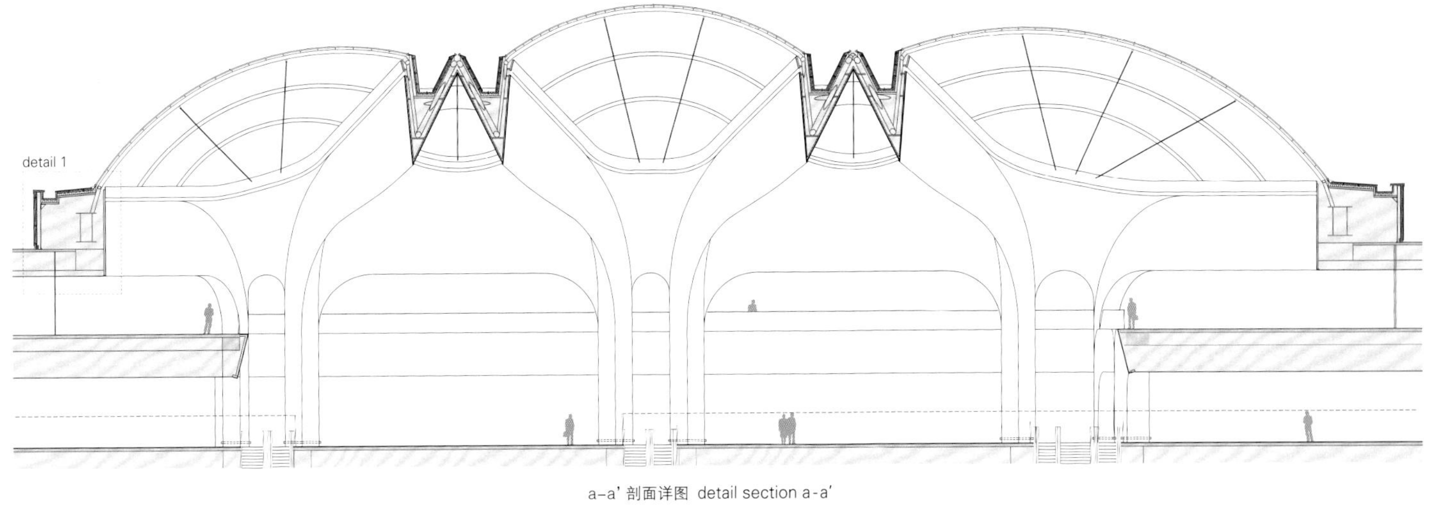

a–a' 剖面详图 detail section a-a'

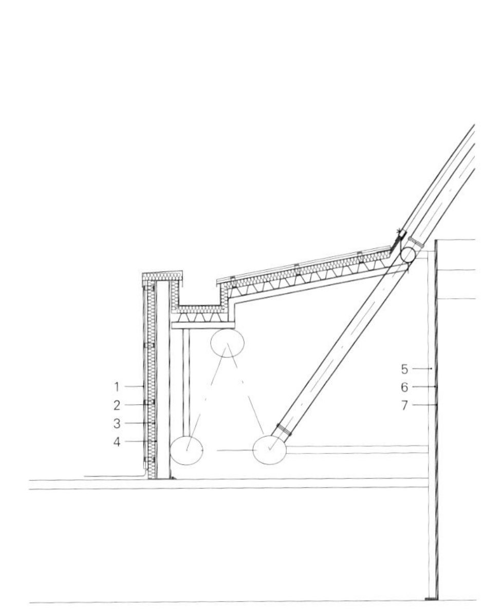

详图1 detail 1

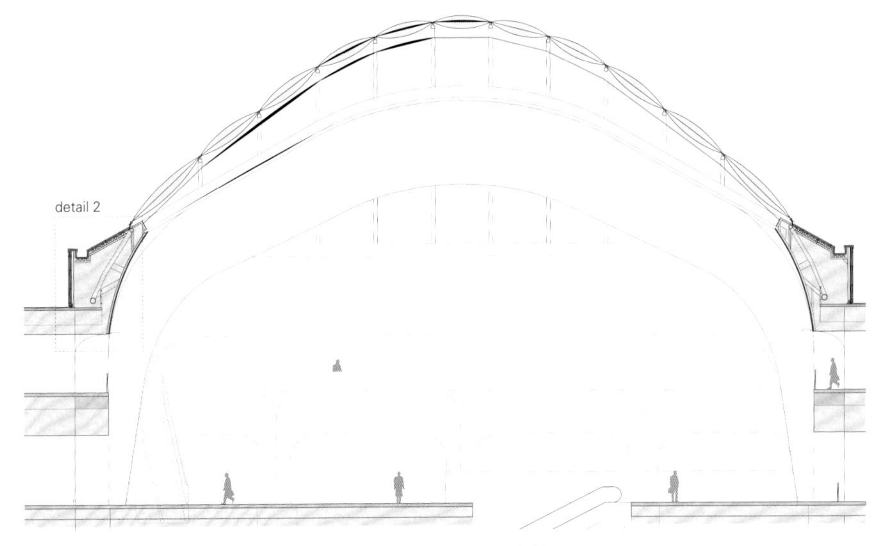

b–b' 剖面详图 detail section b-b'

1. Kalzip cladding 65mm
2. insulation 100mm
3. board 20mm
4. structure zone 200mm
5. metal studwork zone 150mm
6. plaster board 2x12.5mm
7. gypsum based plaster
   -white finish(e.g. Amourcoat perlata)
   -trasparent(water based polyurethane/acrylic) coating to improve the physical properties
8. damper:
   height 475mm; depth 180mm
9. weather louver:
   height 475mm; depth 95mm
   45% penetration(475mmx0.45=215mm)

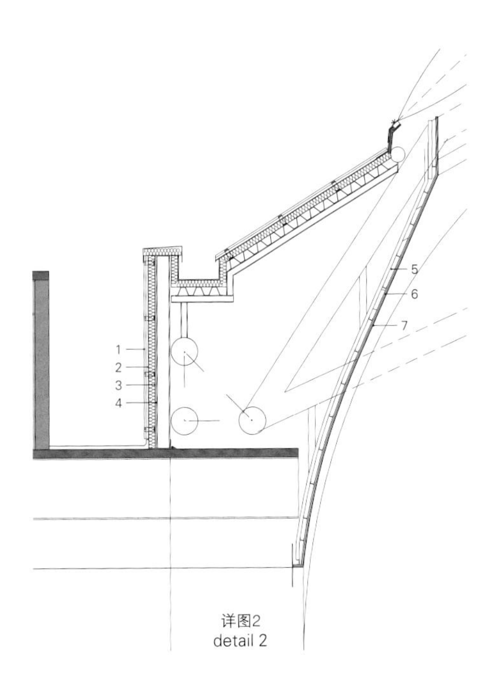

详图2
detail 2

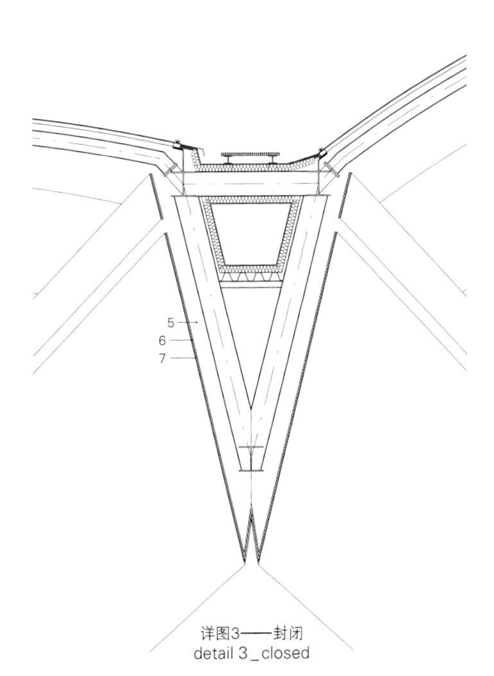

详图3——封闭
detail 3_closed

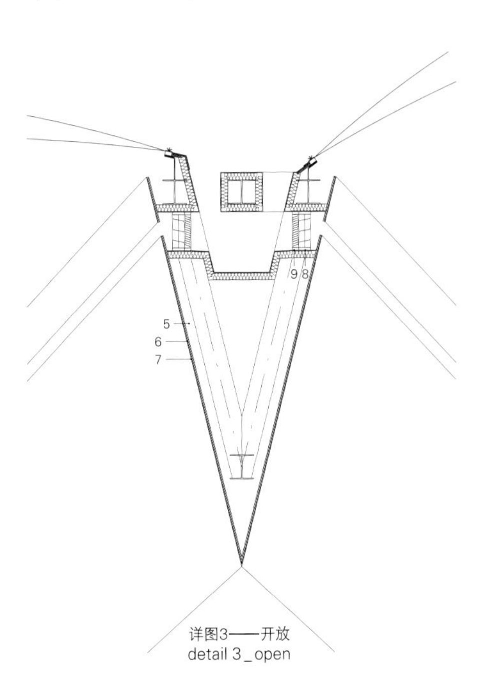

详图3——开放
detail 3_open

利马会议中心

IDOM

利马会议中心项目是秘鲁政府与世界银行、国际货币基金组织之间的协议项目，用于举办利马 2015 年理事会会议。

利马会议中心所处位置极具策略性——位于国家文化中心，毗邻国家博物馆、教育部、国家银行新总部（又称 Huaca San Borja），设计旨在实现四大策略目标：成为国家文化经济发展的助推器；代表植根于秘鲁集体文化的城市核心区的集会地；变成风格独特、功能灵活、技术先进的建筑地标；触发国家文化中心及其周边地区的城市转型。

利马会议中心的净面积接近 15 000m²，共有 18 个多功能会议厅，其面积大小从 100m² 到 3500m² 不等，可容纳多达 10 000 人同时出席各种活动。该项目的其他部分还包括四层地下停车场和一些用作辅助会议室的地上空间。这些空间包括翻译室、会议中心的管理中心、库房、卫生间、工作坊，还有维护保养和物资分发中心、厨房和餐饮区、展览厅、自助餐厅和休息区。整个项目总建筑面积达到 86 000m²。

为了促进文化大道未来的发展，该项目将建筑的入口设在北面。建筑的总体空间格局被设计成三个风格迥异的时空层次，分别象征着国家历史、时间和记忆。

代表现在的是一个被称作"国家房间"的巨大室内开放空间。这儿共有两个可以拆装改变的房间，面积约为 1800m²。但如果把其中一个房间四周的隔声板全部拆掉，它就会向城市完全打开，成为一个面积超过 2500m² 的有顶城市广场。

代表过去的是一个被称作"利马休息室"的室外空间，是本项目的核心，其设计灵感来源于伟大的 huaca（古秘鲁的一种神物）。这一层空间结构由各会议室位置布置及其不同的高度差而自然形成。

代表未来的是一个被称作"国家国际厅"的巨大玻璃体量。这是一个具有高端前沿技术的会议设施，邀请世界各国人民来到秘鲁，提升秘鲁企业家能力，开创秘鲁锦绣未来。

de la Nación

操作的灵活性和功能的多样性是利马会议中心综合设计的重点，定位是达成此项目的经济和社会效益的最大化。由于各个房间是用隔声面板隔断的，空间分配有了多种可能性，所以几乎所有的房间都可以根据需要扩展或缩小。

从技术上来说，设计一个能够容纳 3500 人、面积为 5400m² 的无柱大空间这一强制性条件，再加上使用支撑结构给抗震设计带来的难度，使整个项目无论从设计理念还是结构设计方面都充满挑战，因为这意味着要把这个大空间放到该建筑的最顶层。要在超过 30m 的高度上放置一个相当于一个足球场大小的体量，无论从结构设计还是从建筑内部流动性（进入和撤离）设计方面来说也都是一个挑战。

Lima Convention Center

The Lima Convention Center (LCC) is contextualized by the agreement between the Peruvian State and the World Bank and the International Monetary Fund to hold in Lima the 2015 Board of Governors.

Strategically located in the Cultural Center of the Nation (CCN) – next to the National Museum, the Ministry of Education, the new headquarters of the National Bank or the

Huaca San Borja – the design of the LCC was to satisfy four strategic objectives: being a cultural and economic motor for the country, representing a meeting place at the heart of the city enrooted in the collective Peruvian culture, turning into a unique, flexible and technologically advanced architectonic landmark and finally, triggering the urban transformation of the CCN and its surroundings.

The near 15,000m² of net area correspond to the 18 multi-purpose convention halls, their sizes and proportions varying from 3,500m² to 100m², which allow for up to 10,000 people to attend simultaneous events. The rest of the programme is completed by four underground car-park floors as well as several uses above ground that complement the conference rooms. These would include areas for translation and general management of the centre, stockrooms and toilets, workshops and areas for maintenance and material distribution, kitchens and dining areas, exhibition halls, cafeterias and relaxation areas. This all generates a total built up area of 86,000m².

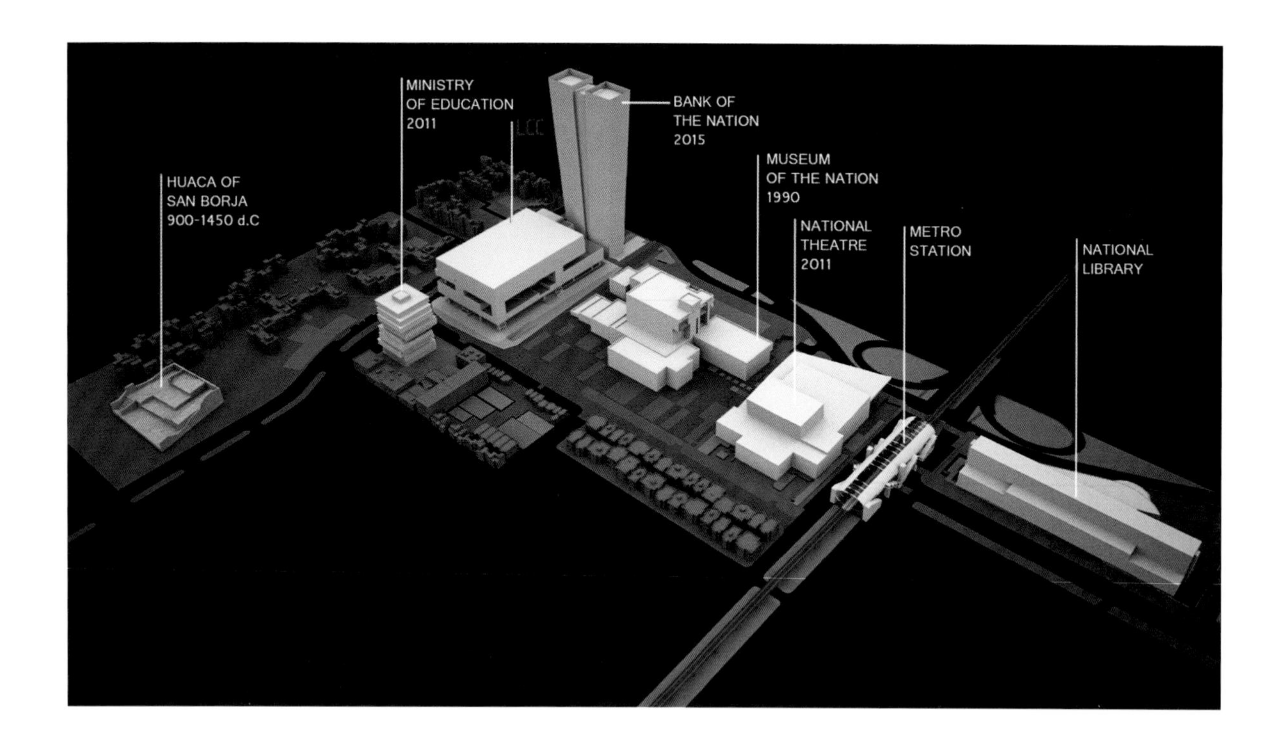
MINISTRY
OF EDUCATION
2011
BANK OF
THE NATION
2015
HUACA OF
SAN BORJA
900-1450 d.C
MUSEUM
OF THE NATION
1990
NATIONAL
THEATRE
2011
METRO
STATION
NATIONAL
LIBRARY

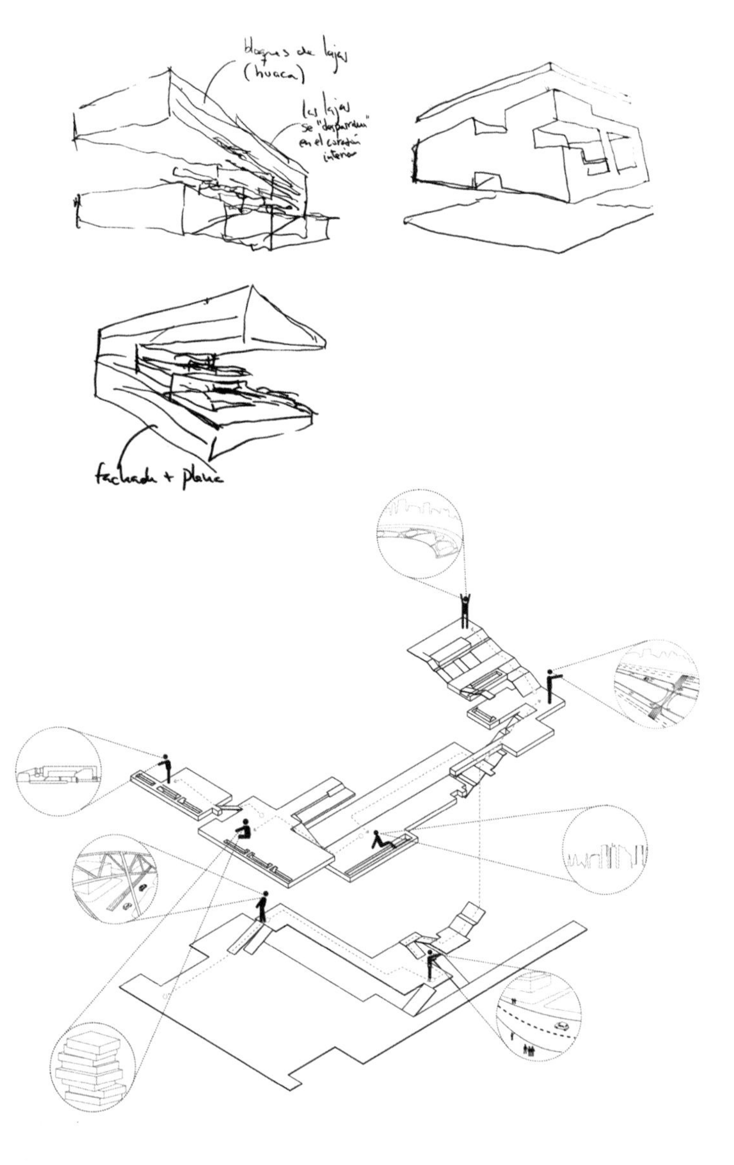

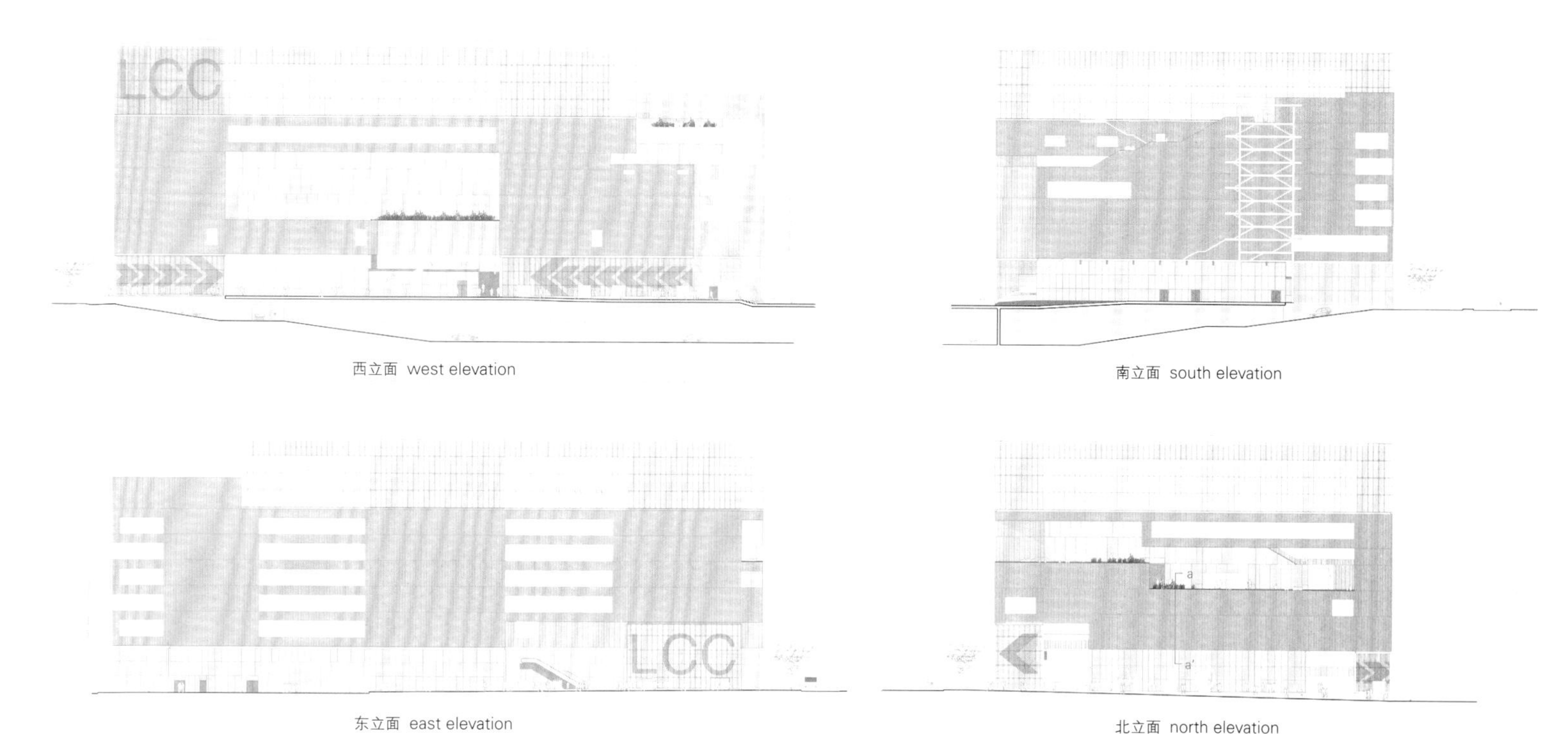

西立面 west elevation

南立面 south elevation

东立面 east elevation

北立面 north elevation

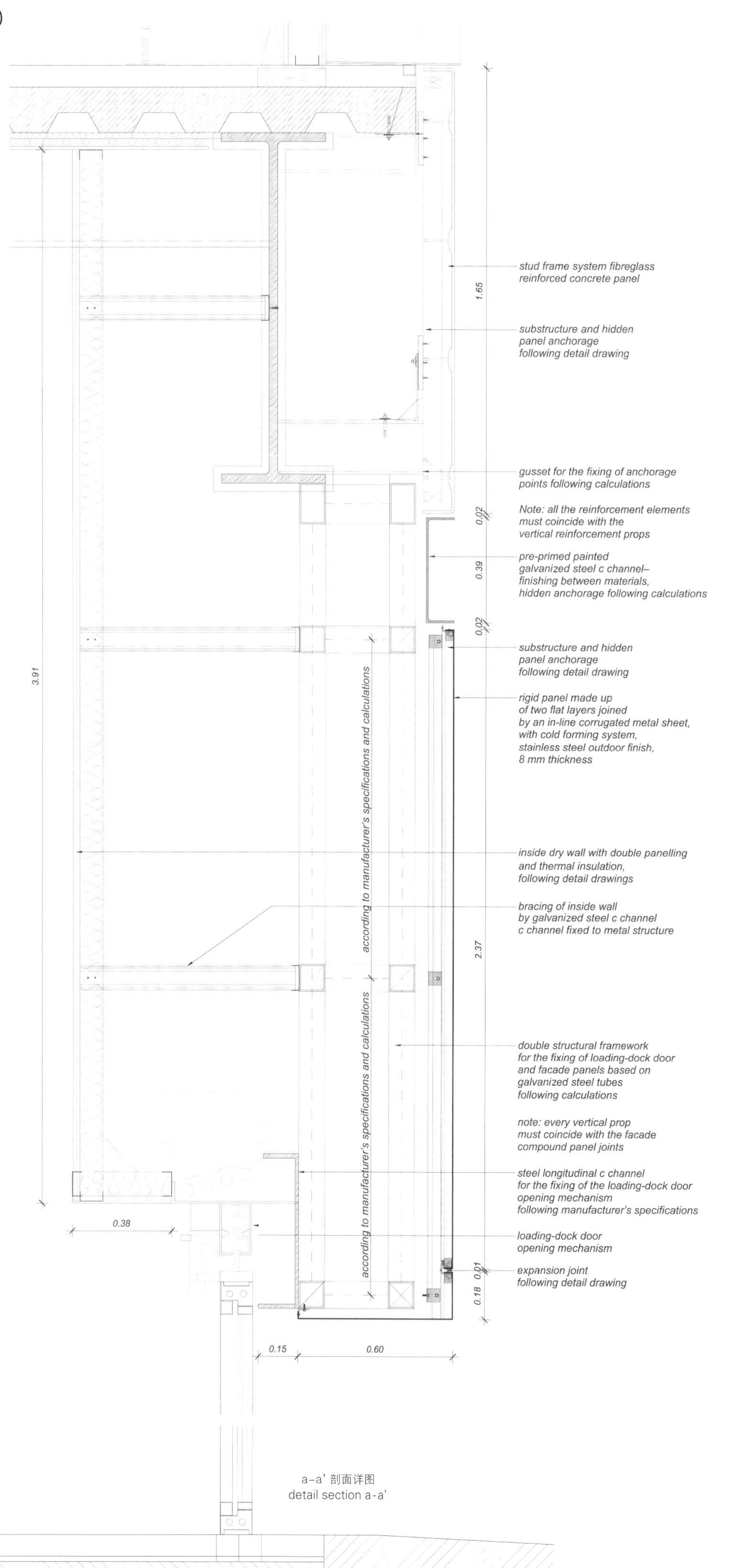
stud frame system fibreglass reinforced concrete panel
substructure and hidden panel anchorage following detail drawing
gusset for the fixing of anchorage points following calculations
Note: all the reinforcement elements must coincide with the vertical reinforcement props
pre-primed painted galvanized steel c channel–finishing between materials, hidden anchorage following calculations
substructure and hidden panel anchorage following detail drawing
rigid panel made up of two flat layers joined by an in-line corrugated metal sheet, with cold forming system, stainless steel outdoor finish, 8 mm thickness
inside dry wall with double panelling and thermal insulation, following detail drawings
bracing of inside wall by galvanized steel c channel c channel fixed to metal structure
double structural framework for the fixing of loading-dock door and facade panels based on galvanized steel tubes following calculations
note: every vertical prop must coincide with the facade compound panel joints
steel longitudinal c channel for the fixing of the loading-dock door opening mechanism following manufacturer's specifications
loading-dock door opening mechanism
expansion joint following detail drawing
according to manufacturer's specifications and calculations
according to manufacturer's specifications and calculations
1.65
0.02
0.39
0.02
2.37
0.01
0.18
3.91
0.38
0.15
0.60

a–a' 剖面详图
detail section a-a'

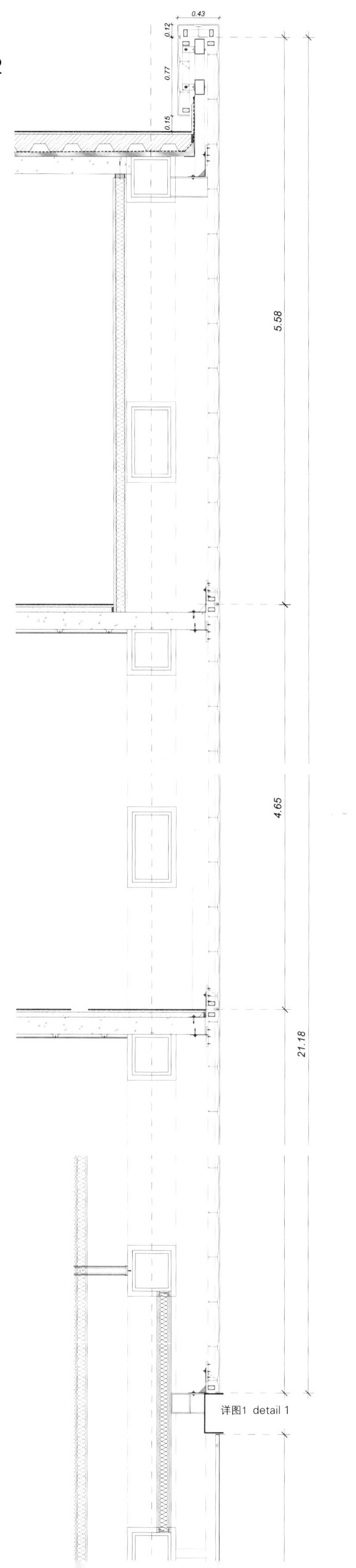

详图1 detail 1

项目名称：Lima Convention Center / 地点：San Borja, Peru
建筑师：IDOM / 项目主管：Javier Álvarez de Tomás
建筑师负责人：Tono Fernández Usón, César Azcárate, Javier Álvarez de Tomás
项目管理：Javier Álvarez de Tomás, Miguel de Diego, Alejandro Puerta, Carmen Camarmo / 项目团队：María Cortés, Enrique Alonso, Jorge Rodríguez, Alejandra Muelas, Adrián Jabonero, Roberto Moraga, Armide González, Nazareth Gutiérrez, Mª Amparo González, Lucía Chamorro, Jesús Barranco, Jesús Llamazares, Borja Gómez, Juan Pablo Porta, Pablo Viña, Luis Valverde
结构工程师：Alejandro Bernabéu_General coordination, Javier Gómez, Mónica Latorre /外围护结构：Magdalena Ostornol (Skinarq)
设施：Antonio Villanueva, Ramón Gutiérrez, Ulises Rubio, José Antonio Yubero_electricidad, telecomunicaciones, José Manuel Jorge, Carlos Jiménez, Mariano Traver, Luis Martín, Celia Monge (Solventa), Javier Martínez (Solventa)
声学设计：Mario Torices / 照明设计：Noemi Barbero
施工建筑建议：Tono Fernández Usón, Miguel de Diego, Javier Álvarez de Tomás / 客户：Constructora OAS, Sucursal del Perú
用地面积：10,617m² / 建筑面积：86,715m²
设计时间：2014 / 施工时间：2014—2015 / 竣工时间：2016
摄影师：©Aitor Ortiz (courtesy of the architect)

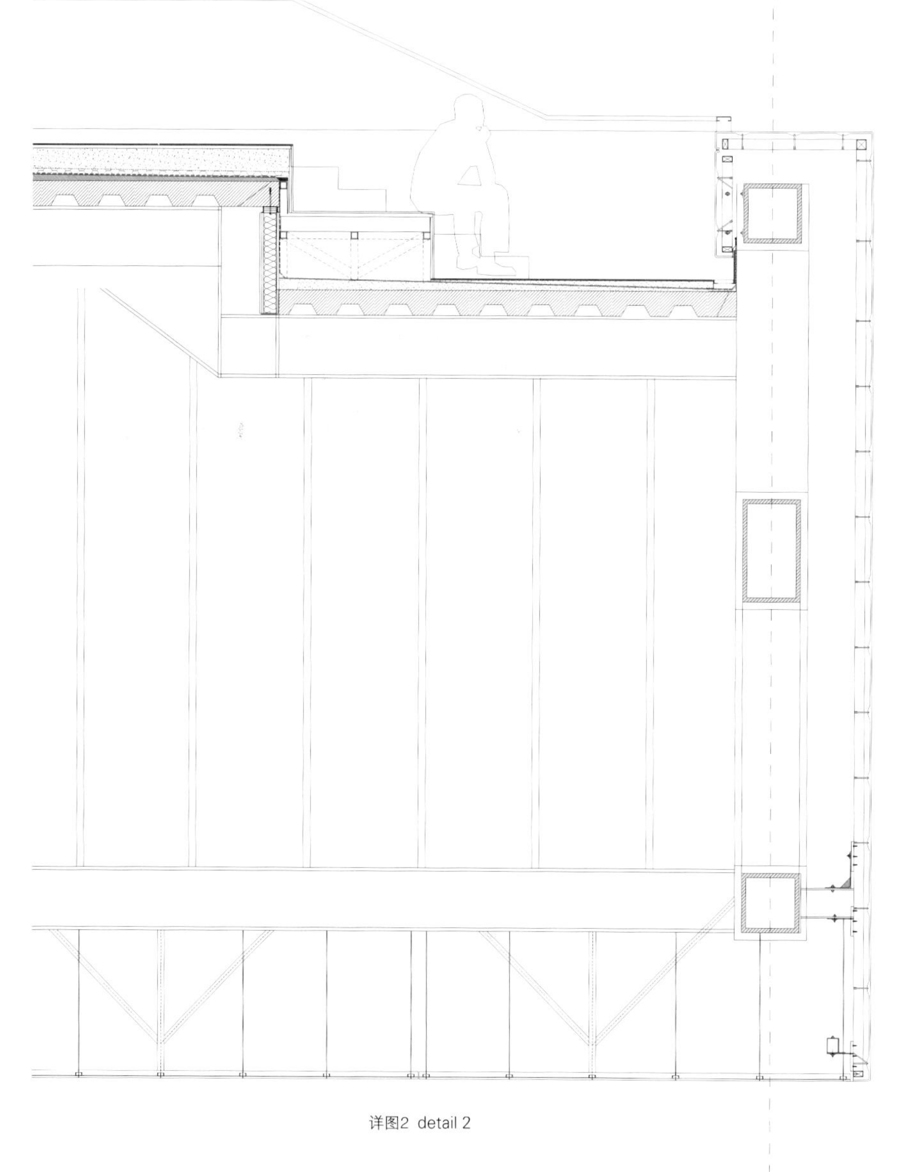

详图2 detail 2

1.00

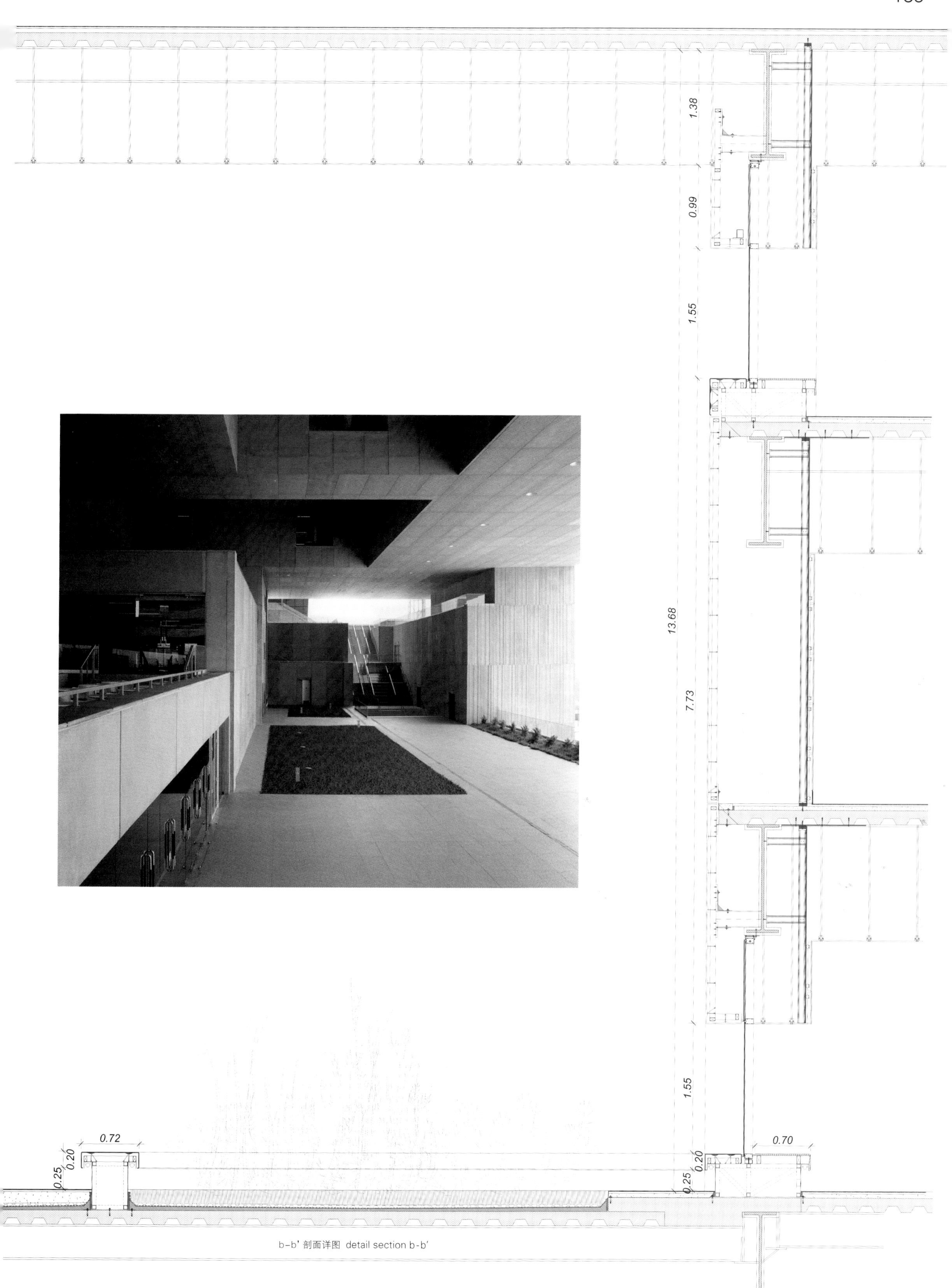

b-b' 剖面详图 detail section b-b'

The urban proposal situates the access to the building on the north end, therefore encouraging the future development of the Culture Boulevard. The general volume is organized into three time-physical strata clearly differentiated, symbolically related to the country's history, time and memory:
The present is represented by the great internal void – Nation Rooms – which harbours the two transformable rooms of about 1,800m², one of which can open up entirely to the city by clearing its perimeter of the acoustics panels that make it up, generating a sheltered urban plaza over 2,500m².
The past, the heart of the project, is an outdoor area inspired by a great huaca – Lima Lounge – generated naturally by the disposition and the difference in height of the convention halls.
The future is a great vitreous volume – International Room of Nations. It's a highly technical conventions facility which invites the rest of the world to come to Peru for its entrepre-

neurial capacity and its promising future.
The operative and functional flexibility are keys to the comprehensive design of the LCC and are orientated towards maximizing the economic and social success of the project. Nearly all rooms can be extended or reduced thanks to the acoustic panels that limit them, making it possible to have several spatial distributions.
Technically, the mandatory condition by which the great 5,400m² room, with capacity for 3,500 people, was to be free from pillars – along with the seismic inconvenience of using propped up structures – turns the conceptual and structural proposal into a challenge, since it implies putting the great room on the last level. Placing a sheltered volume the size of a football pitch at a height of over 30m is a challenge to both the structural approach and the building's internal mobility – access and evacuation.

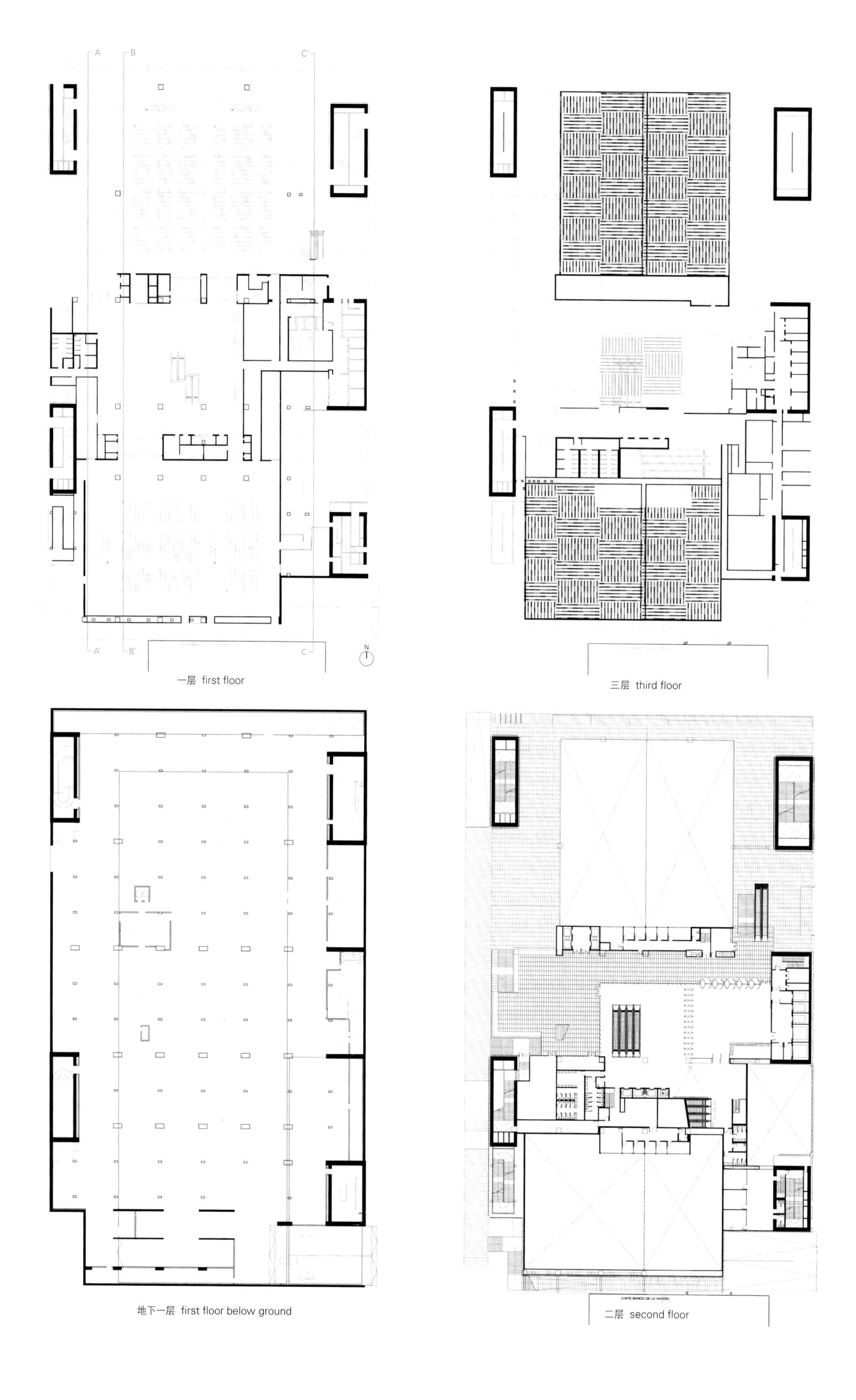

一层 first floor

三层 third floor

地下一层 first floor below ground

二层 second floor

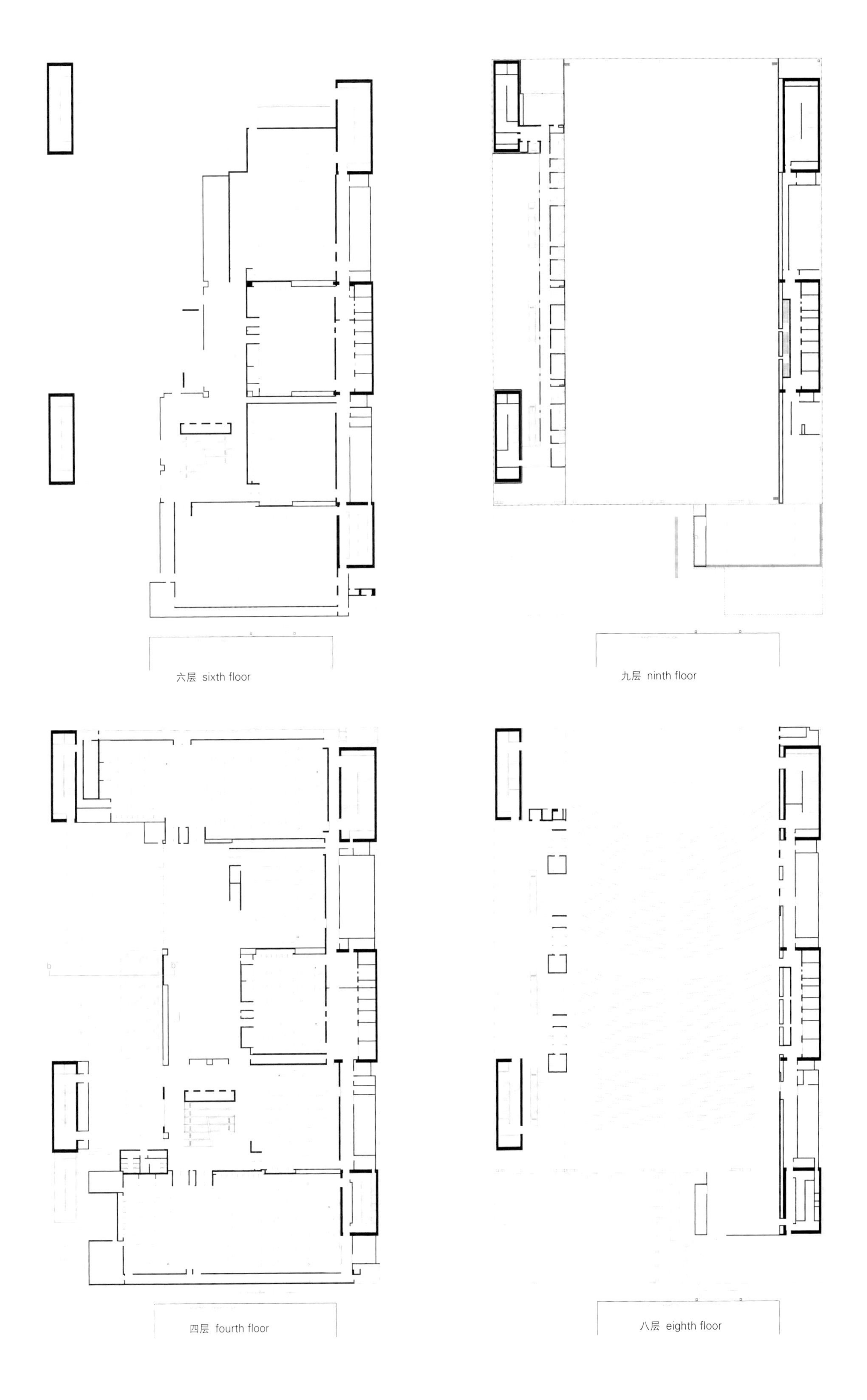

六层 sixth floor

九层 ninth floor

四层 fourth floor

八层 eighth floor

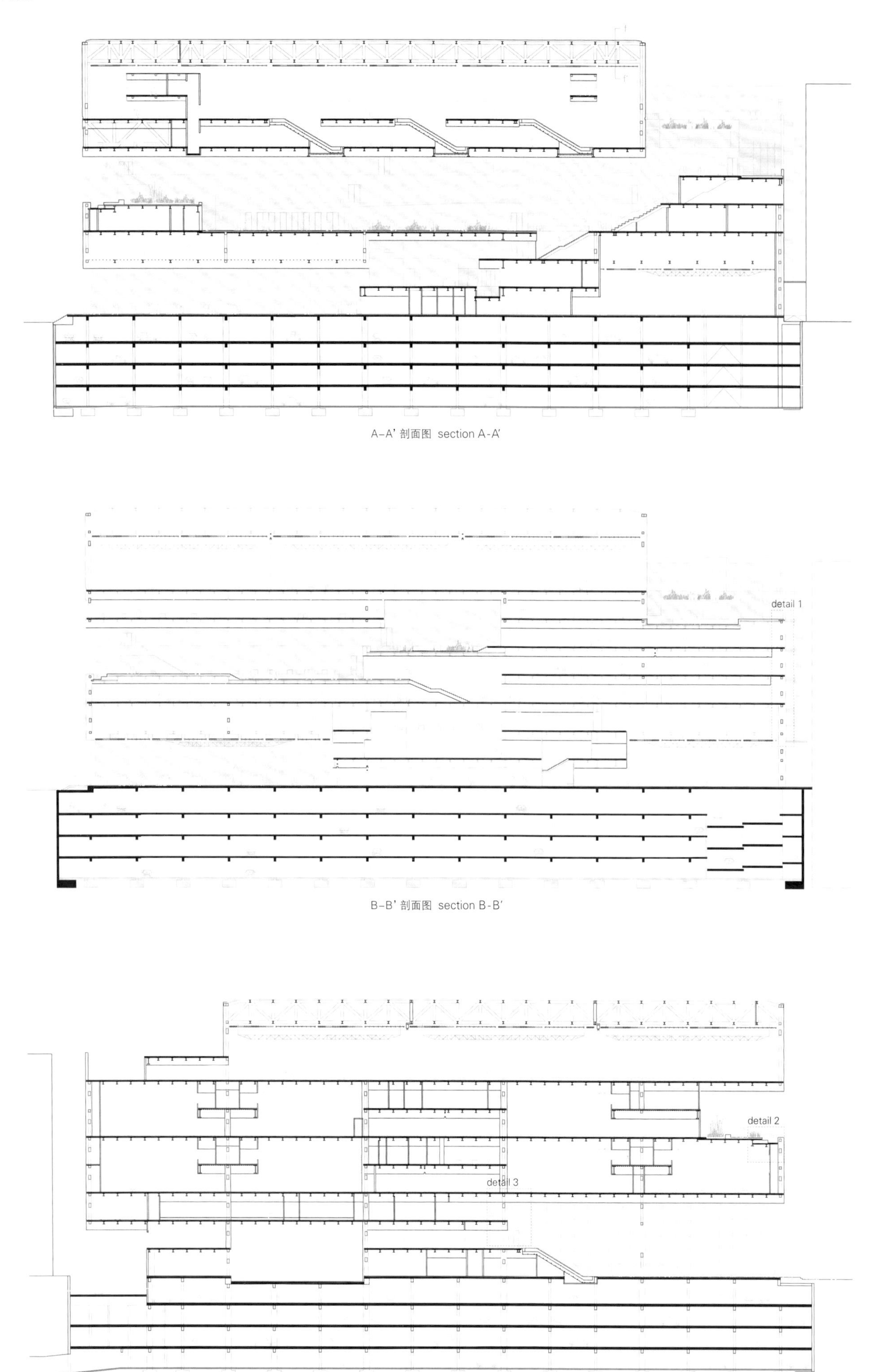

A–A' 剖面图 section A-A'

B–B' 剖面图 section B-B'

C–C' 剖面图 section C-C'

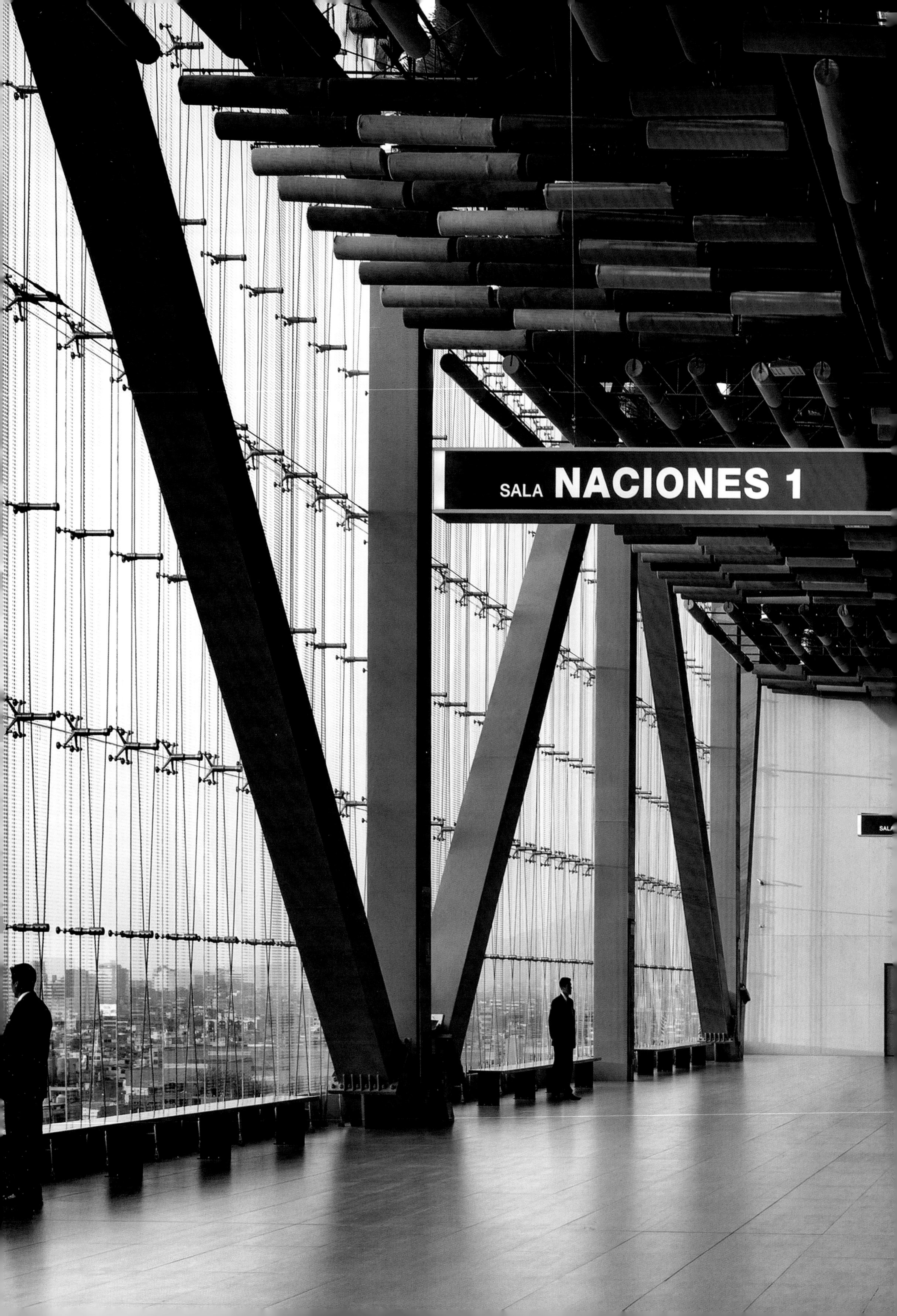
SALA NACIONES 1

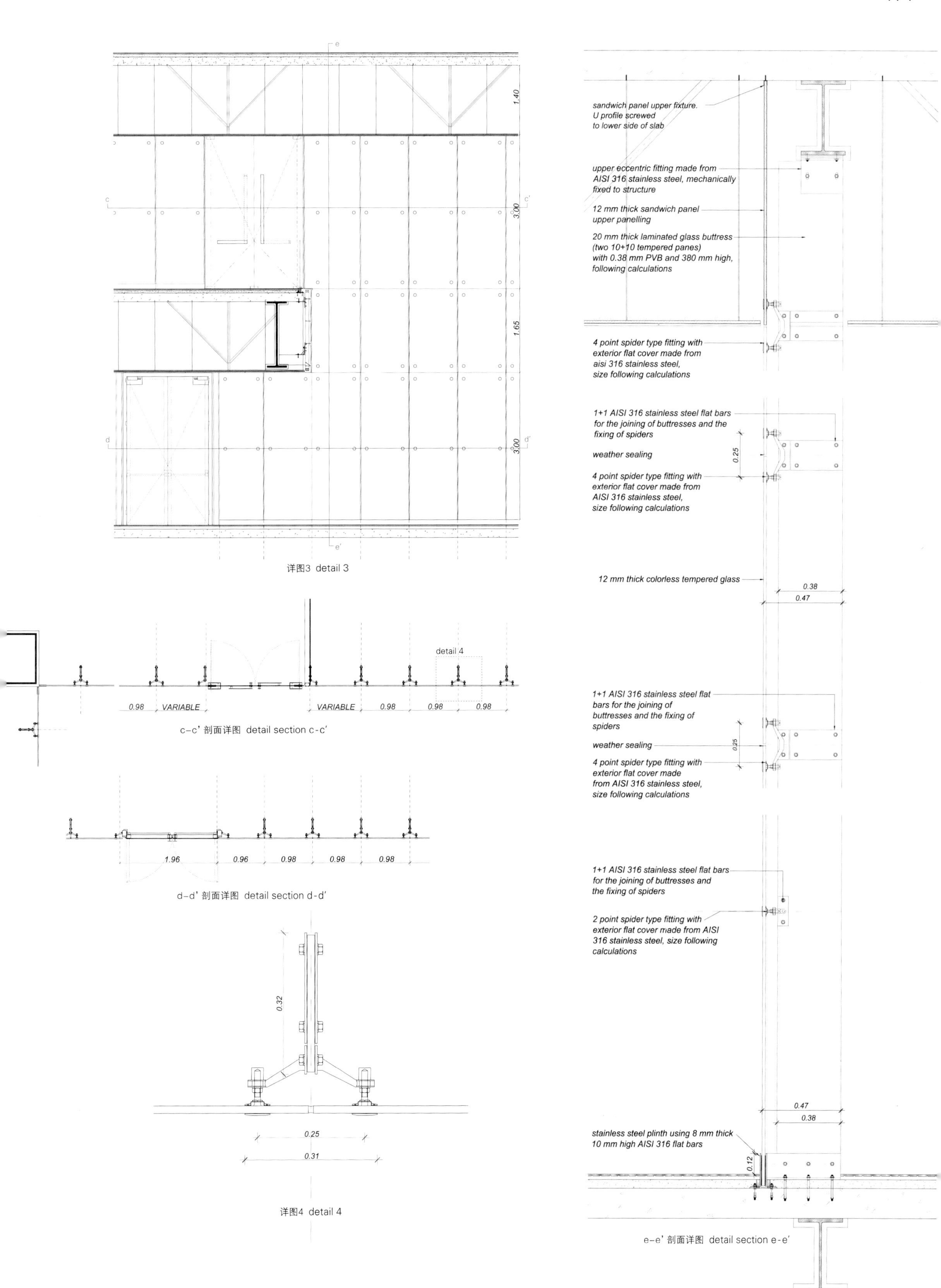

详图3 detail 3

c–c' 剖面详图 detail section c-c'

d–d' 剖面详图 detail section d-d'

详图4 detail 4

e–e' 剖面详图 detail section e-e'

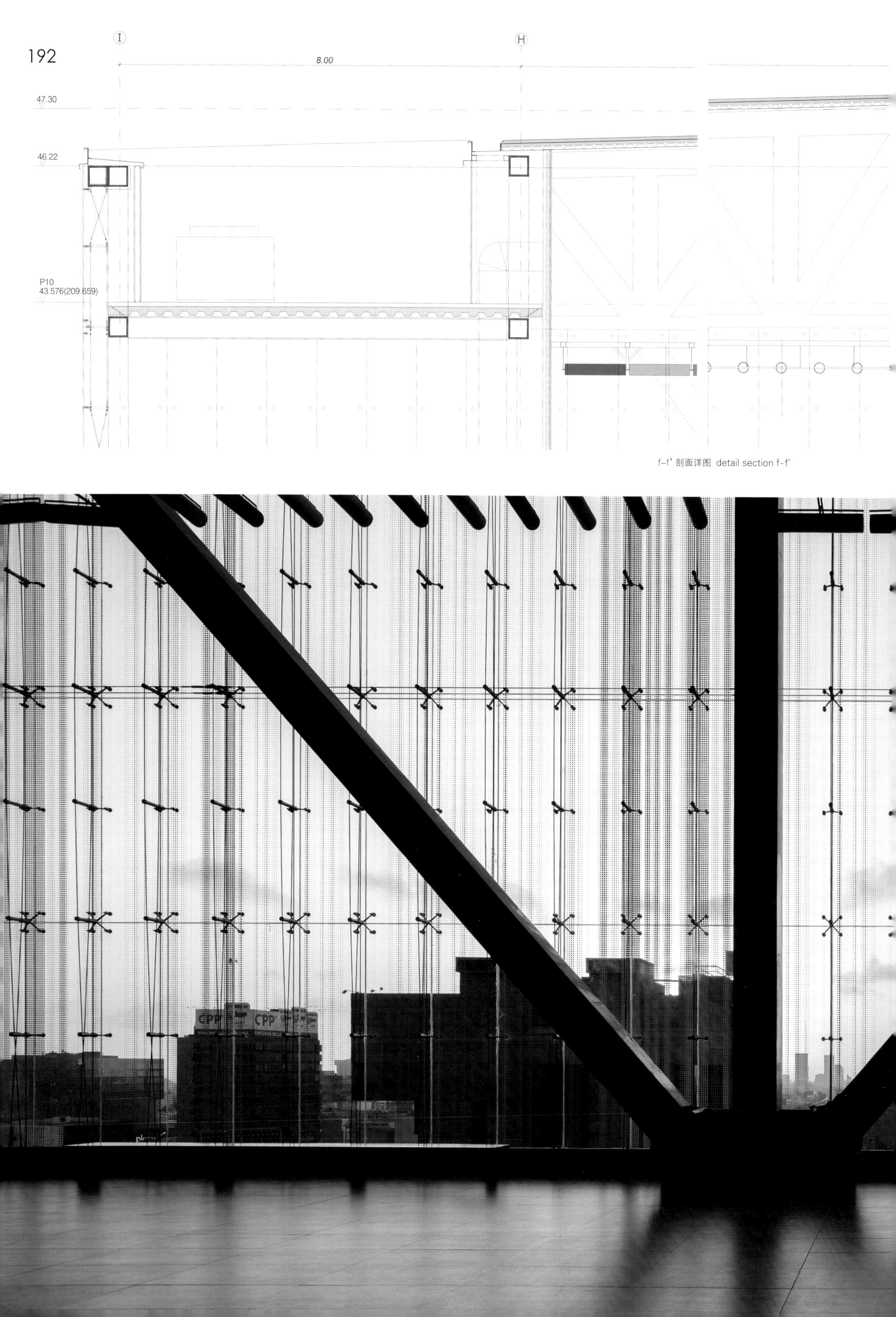

f–f' 剖面详图 detail section f-f'

B
A
10.00
5.50
SALA NACIONES 2

帕萨亚回力球场与公园

VAUMM

Pasajes 湾的自然地理位置得天独厚，其开口长达 300m，两侧是悬崖峭壁，自然形成了一条港湾通道，保护着里面的船只，而港湾里面则成为一个无与伦比的船只避风港。

多年以来，这片土地的历史进程已经证明，其最大的优点也是其最大的问题。工业革命期间，Pasajes 湾海港所拥有的优异海洋条件带来了经济成倍的增长，将其独特的自然景观变成了一派生产忙碌的景象。一方面，这象征着该地区工业的发展；另一方面，象征着人们改造自然的力量。重金属加工业、大型铁路交通基础设施、煤电火力发电厂和无数的物流公司在这个独特的自然环境中安家落户，完全改变了该地区的外观和经济。

随着从重工业模式到技术型产业模式的转变，人们具备了构建新社会的理念，意识到重塑环境的必要性，因而现在释放出许多地块，准备将这些地块改造成居民区。

拆除废弃的工业建筑，对这些建筑曾占用的土地加以去污净化，结果使得土地变得如同月球表面一样坑坑洼洼，看不到任何过去的痕迹，也没有可辨认的地理形态，更没有明显的参照物，只有明显定义此地边界的铁路基础设施、环城公路和 Molinao 河。

在这个像"白板"一样的地区，首先要做的是建立起一个新的地理形态。利用已经挖掘好的空地来修建一座回力球场，其外形就像一个山丘。球场边一条绿茵茵的绿化带将这个新建的城市区与环形公路隔离开来。同样起到隔离作用的还有一条沿着场地高起的一侧设计的一条运动跑道。跑道两头与原来的小路相连，其边上还有一个新平台，有一些供老年人使用的运动器材。

场地的中间区域被设计成一个公共广场，与回力球场相连，是典型的巴斯克回力球场设计风格。在这种设计中，广场与球场共同形成一个不可分割的单元。广场可以空无一物，人们在此可以举办各种活动，如音乐会、午餐会、地区性马戏巡回表演等。回力球场也保持着它的多功能性，放弃了它作为球场的专业性。在它 36m×14m×12m 的三面体空间中，可以举办各种文化和社会活动，朝向它的露天看台可容纳 250 名观众。

TOYOTA

在地形改造设计中，最后一个层次是沿 Molinao 河设计的一条人行步道。为了匹配河岸两边的高度，建筑师设计了预制混凝土墙。预制混凝土墙让人一眼就能看出原始的高度，新的城市景观就建于其上。

球场是整个设计的重点。球场的面积要求建筑的体量很大，但为了不破坏周围环境，建筑师通过掩埋部分体量使其拥有一个合适的规模。缩小规模，再加上像对待一件物体一样精心处理，建筑师的设计使球场建筑看上去更像一个雕塑。还有一个小型咖啡亭和一个自动卫生间，这两个功能设施同时为回力球场和广场的公众服务。

这种多面几何体的建筑总是给人们呈现出雕塑体量碎片化的一面。从不同角度去观察，它的形态都是不一样的。这一点在建造过程中被再次强调，建筑每一面上锌板排列的方向都不尽相同。同样的不规则形态也体现在其他构件上，如扶手、大门，甚至花池和凳子等，它们跟建筑一起构成了一个完整的整体。

Pelota Court & Park in Pasaia

Pasajes Bay is a privileged natural location, determined by a mouth of 300 meters long; a natural corridor flanked by cliffs, which gives access to the seaport and protects the ships inside, setting an exceptional refuge seaport.

The historical evolution of this territory has verified over the years, that its great virtue was its great problem, indeed. The magnificent sea conditions of its harbor which brought it on exponentially, during the Industrial Revolution, transformed this exceptional natural territory into a productive landscape.

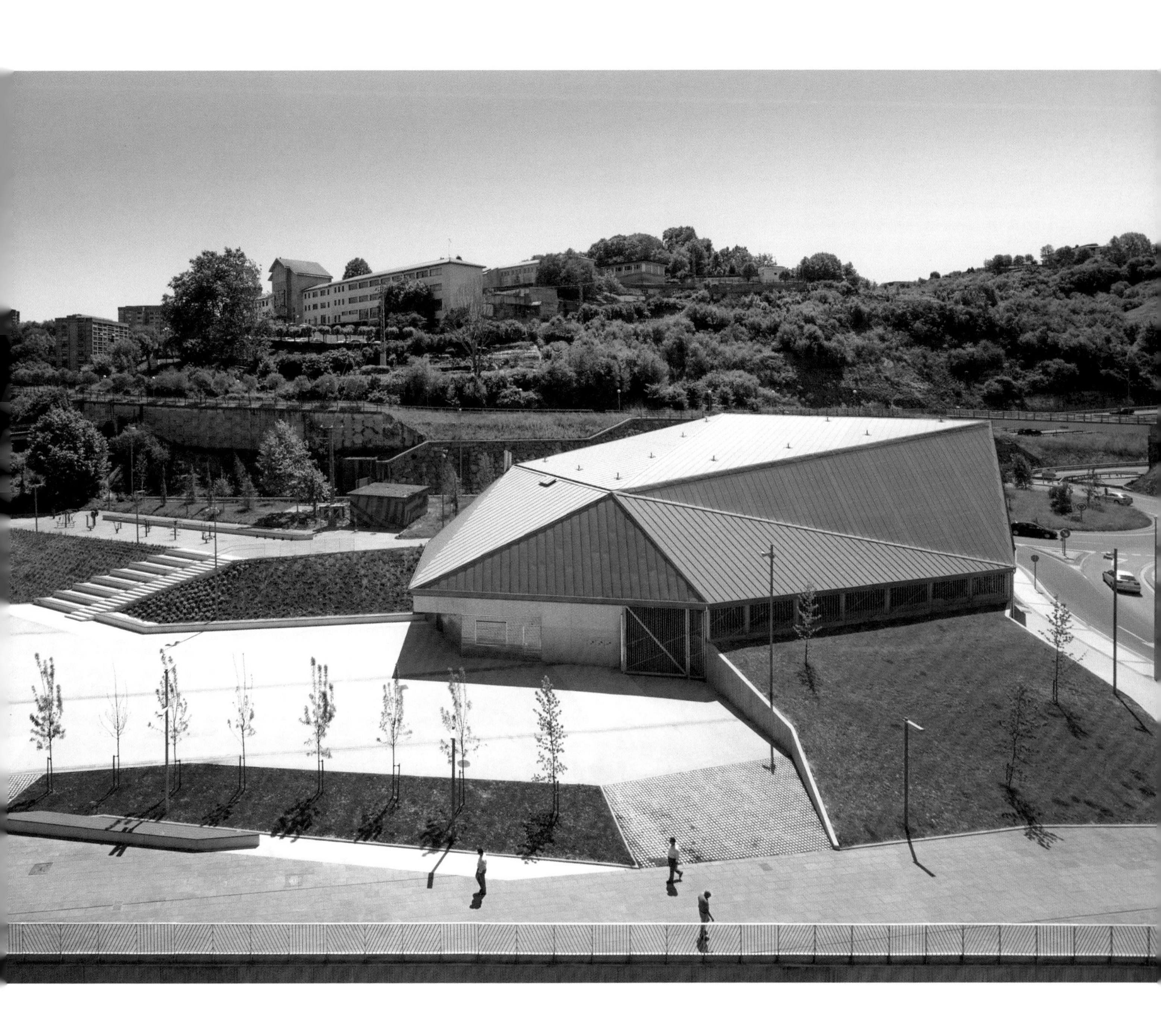

On one hand, emblem, of the industrial development of the region, on the other hand, emblem of the transforming power of Man. In this unique natural setting, heavy metal processing industries, great rail transport infrastructure, coal based thermal power plant and countless logistic industries were settled down, completely transforming both appearance and economy of the region.

The evolution of the productive system, from heavy to technological industry and the logics of a new Society, aware of the need to restore the Environment, released many plots, which now are transformed to serve residential areas.

The dismantling of abandoned industrial buildings and the decontamination of the soils occupied by these buildings, left as a result, a lunar terrain, with no references to the past, with no recognizable topography, with no significant reference points, apart from its clearly defined edges: railway infrastructure, ring road and Molinao river.

In this place, result of "tabula rasa", the first intervention was setting up a new topography. Taking advantage of the required excavation to build the Pelota Court, a small hill is given shape, which besides is complemented with a powerful wooded dough; this way it's built a spacer element for this

北立面 north elevation

南立面 south elevation

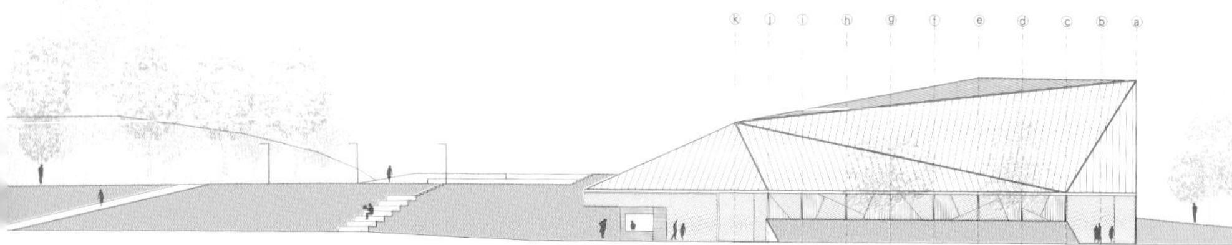

西立面 west elevation

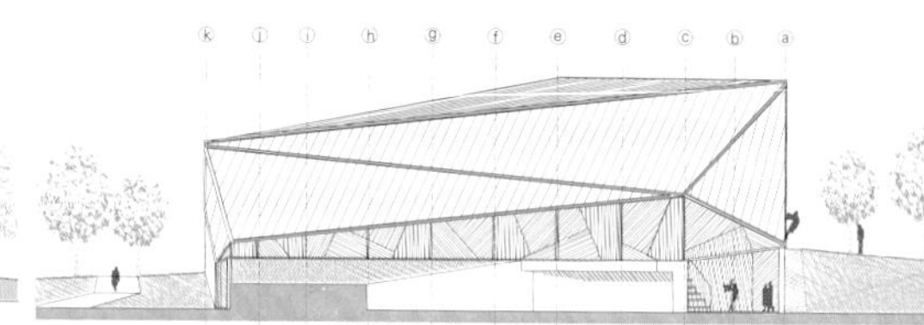

西南立面 south-west elevation

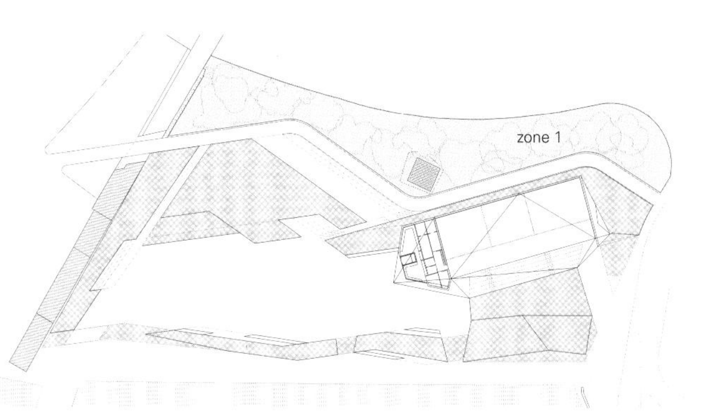

花园——区域1
gardening_zone 1

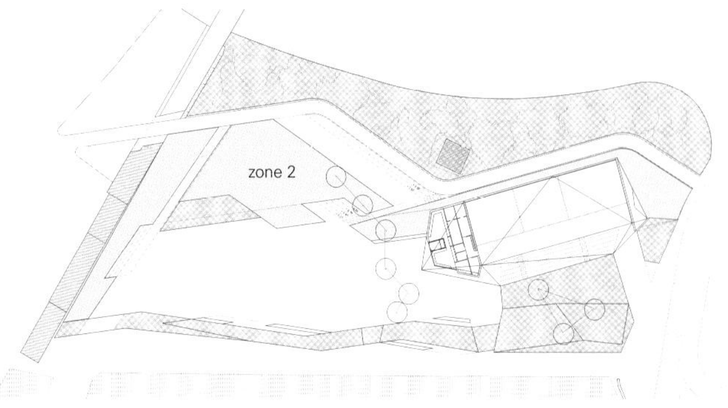

花园——区域2
gardening_zone 2

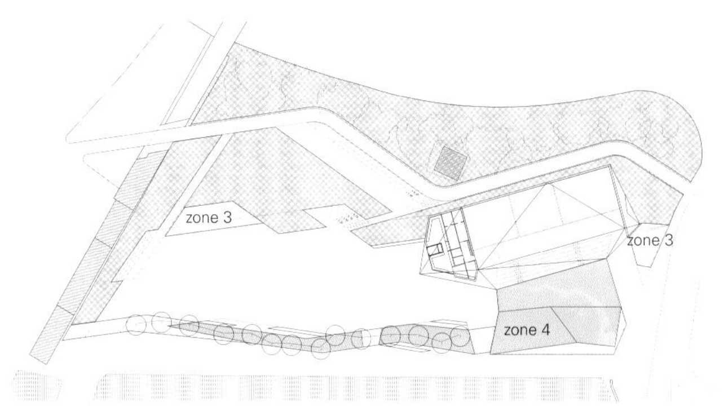

花园——区域3&4
gardening_zone 3 & 4

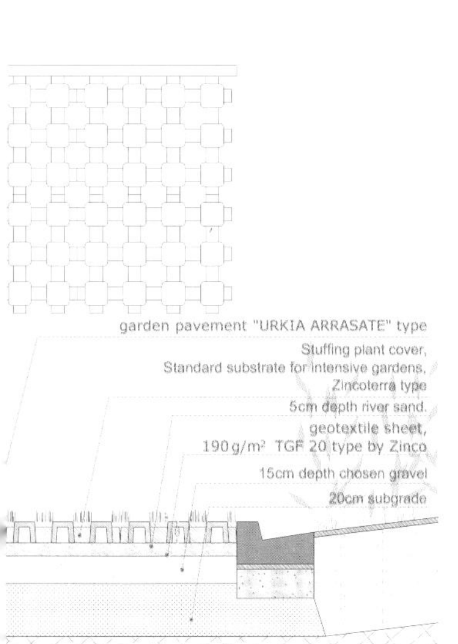

混合铺地材料与绿色斜坡详图
mixed paving and green slope detail

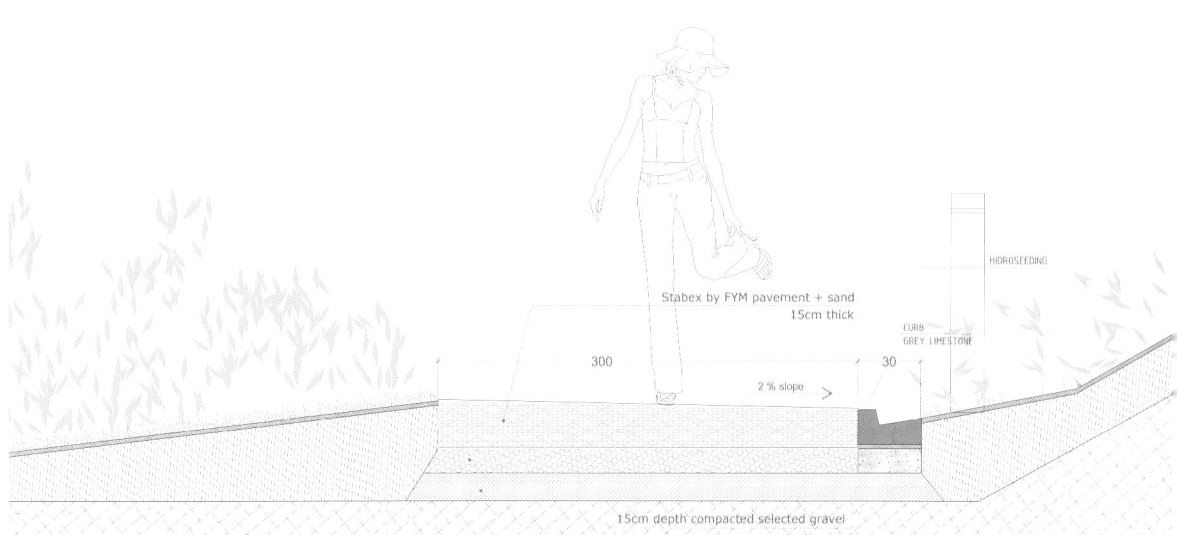

高处道路和绿色斜坡详图
high path and green slope detail

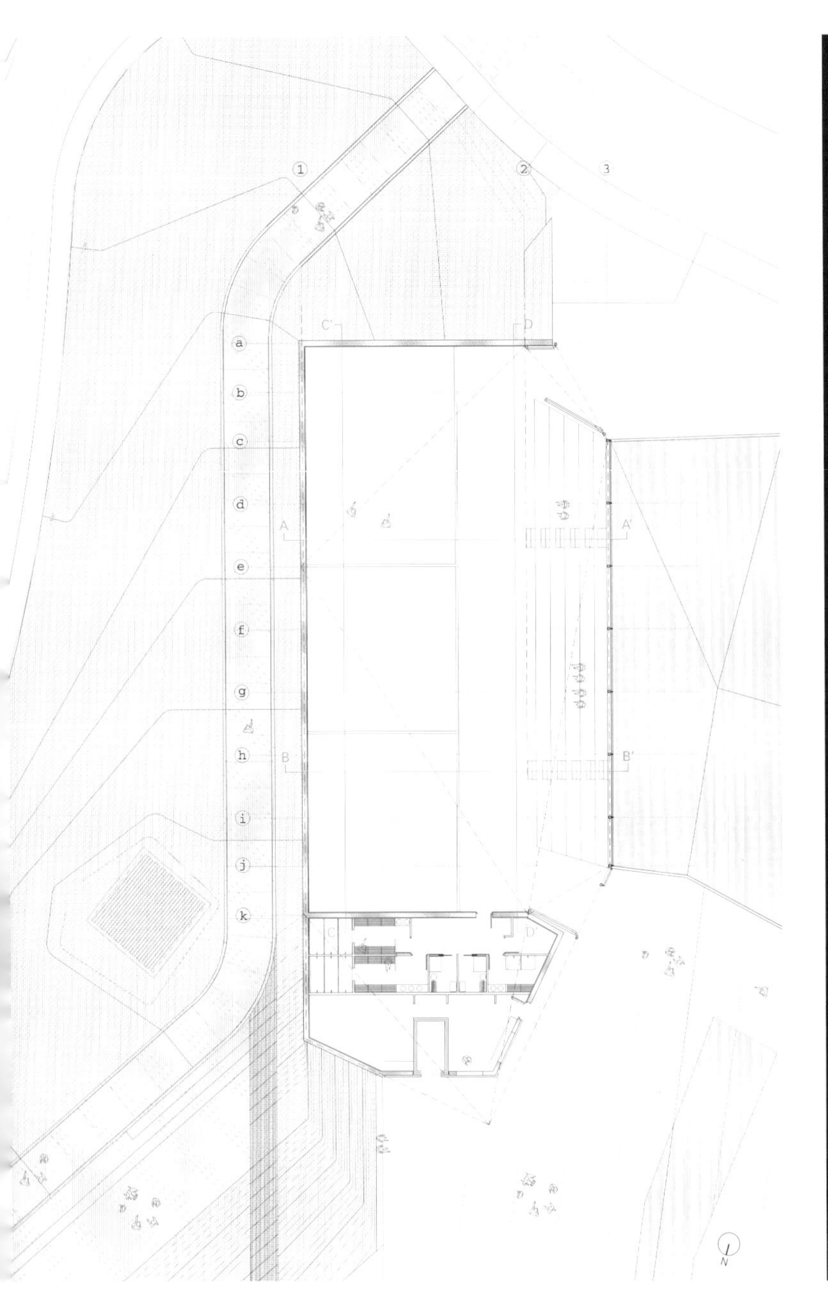

new urban area, from the ring road. This element is further supported by the construction of a sports path that runs along the highest point of the site, connecting with both ends of existing paths, which besides serves as a support to a new platform with sports sets for elderly people.

The intermediate area of the place is configured by a public square, linked to Pelota Court, recalling the classic urban typology of Basque Pelota Courts, where Square and Pelota Court form an indivisible unit. The square is configured as an "empty" space, where it's possible to take place a variety of events, from concerts, popular lunches, to circus on regional tour. The Pelota Court also maintains its multipurpose spirit, and gives up its regulatory measures of the game of Pelota, for all kind of cultural and social events that can take place in the emptiness of its trihedron of 36 x 14 x 12 meters, open to bleachers for a capacity of 250 spectators.

The last layer of the intervention sets a riverside promenade accompanying the Molinao river. The need to match the levels of both edges of the riversides is solved by a prefabricated concrete wall, revealing at first glance the original level, on

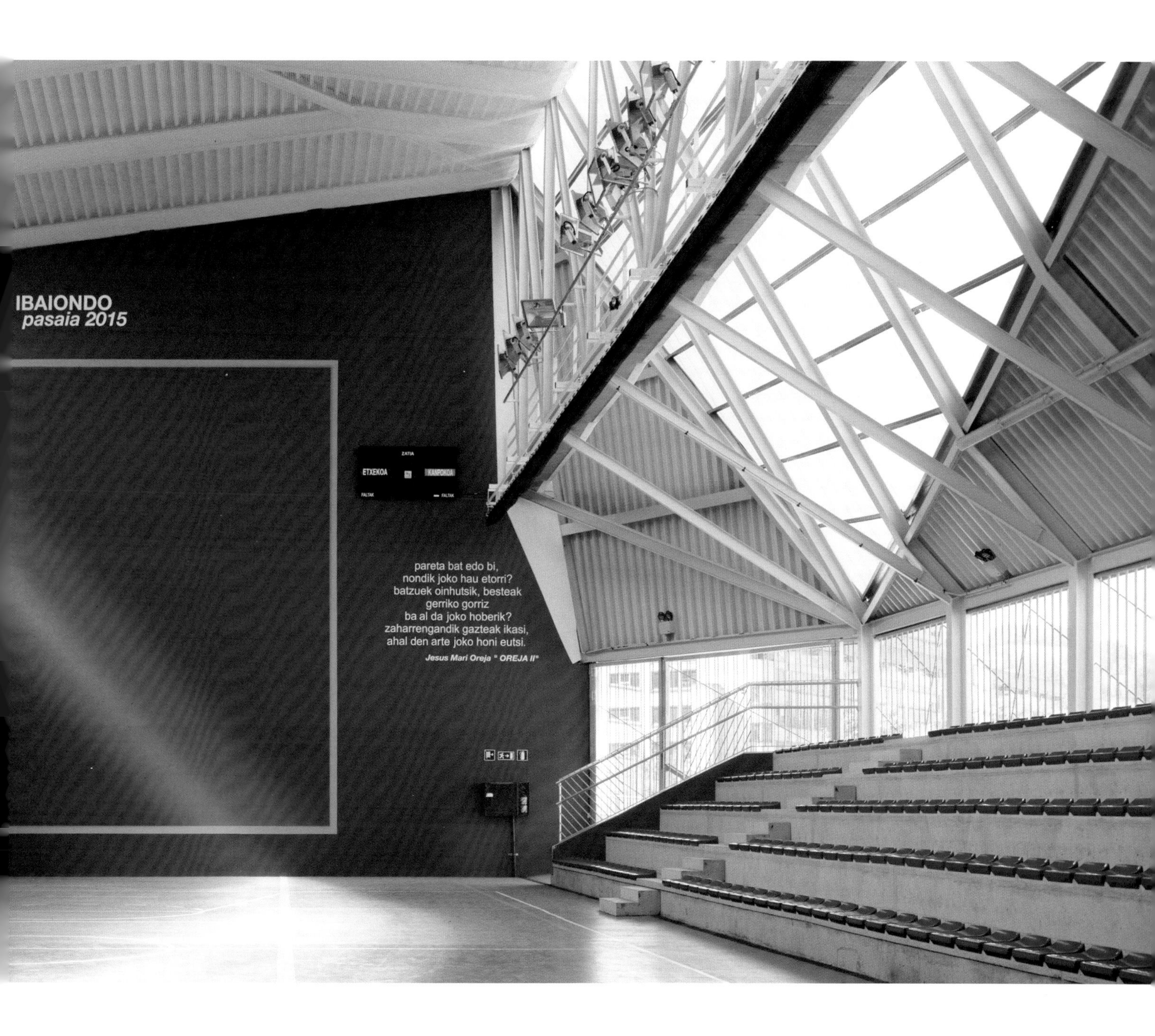

which the new urban landscape rises.

The built object of the Pelota Court is the most prominent element of the intervention. Its dimensions require a large volume, which through its partial burial, gets a suitable scale, respectful to the surroundings. This downscaling is emphasized by the treatment of the built piece as an object, turning the building into an almost sculptural element. A little "kiosk" cafe and an automatic toilet, serve as intermediate functional elements between the uses of the Pelota Court and the square.

The faceted shaped geometry of the building, always suggests a fragmented reading of the sculptural volume of the piece. Its presence, always changing, is the result to several points of view. This same planes key reading is amplified due to the construction, since the different sides impose different orientations for each zinc sheet, providing a texture that singularly designs each plane. These same patterns of faceted shaped geometries organize the proper elements of urbanization, such as handrails and closings, as well as flowerbeds or seat banks, building through geometry and matter, a unity of different elements that form a whole.

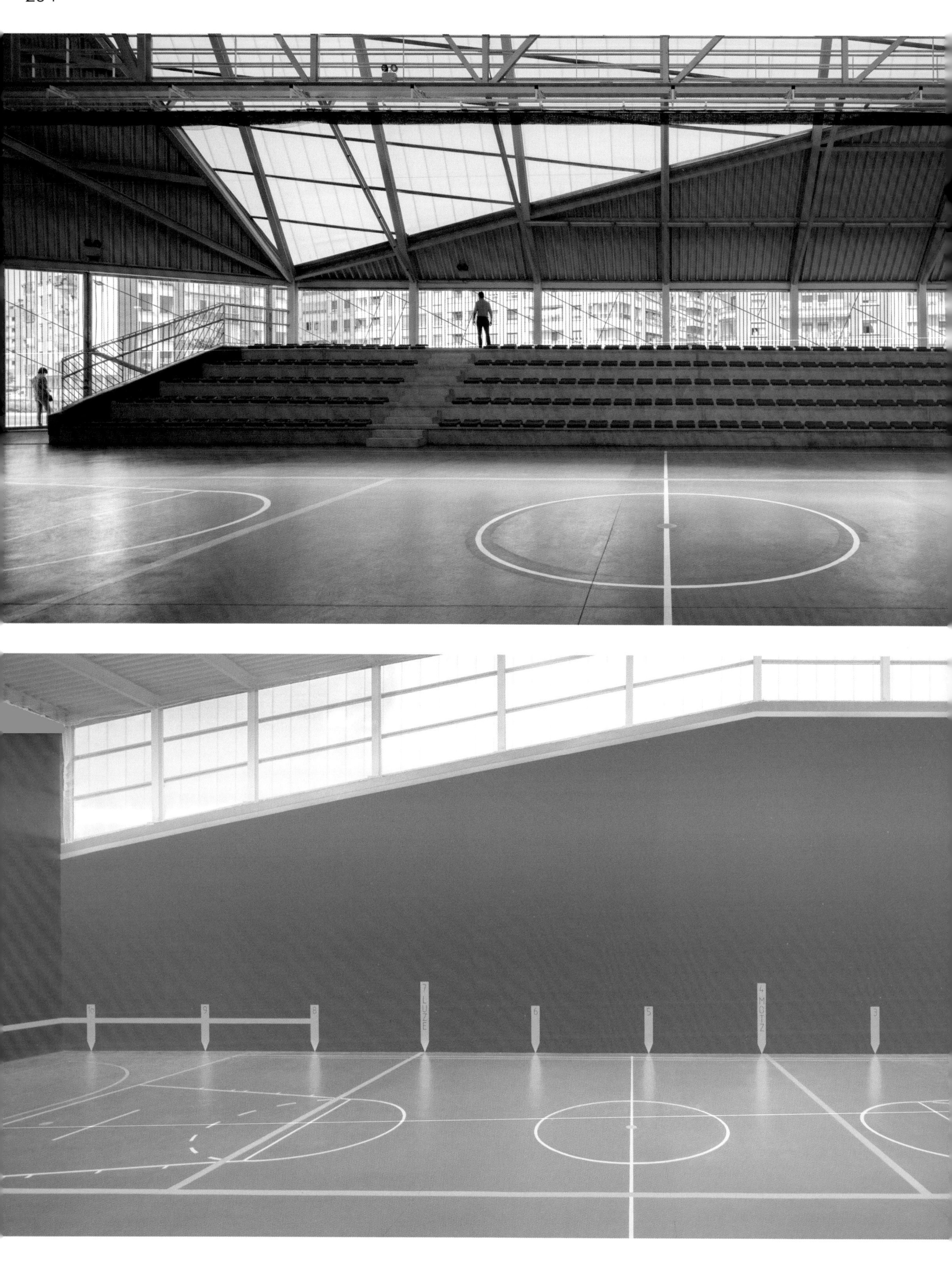
10
9
8
7 LUZE
6
5
4 MOTZ
3

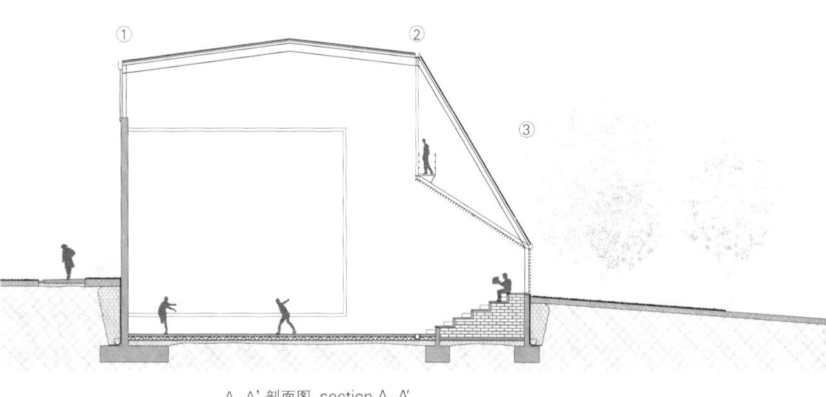

A–A' 剖面图 section A-A'

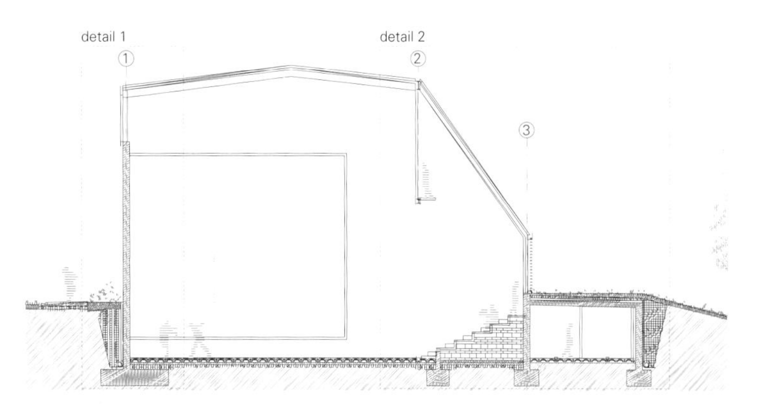

B–B' 剖面图 section B-B'

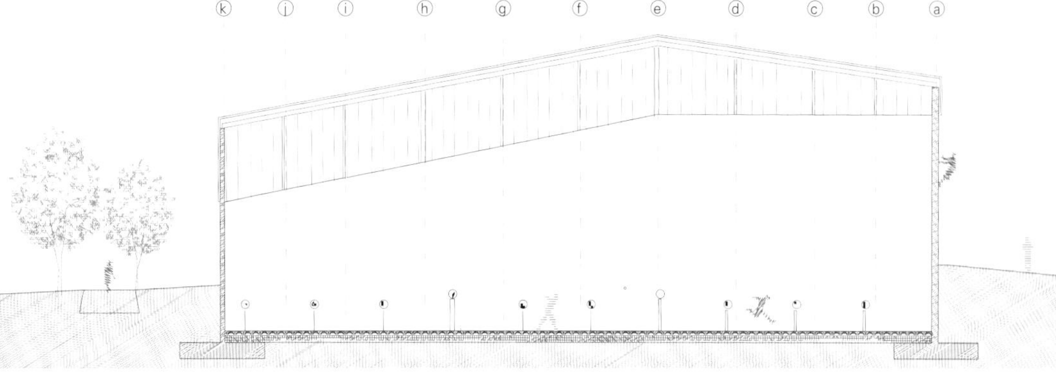

C–C' 剖面图 section C-C'

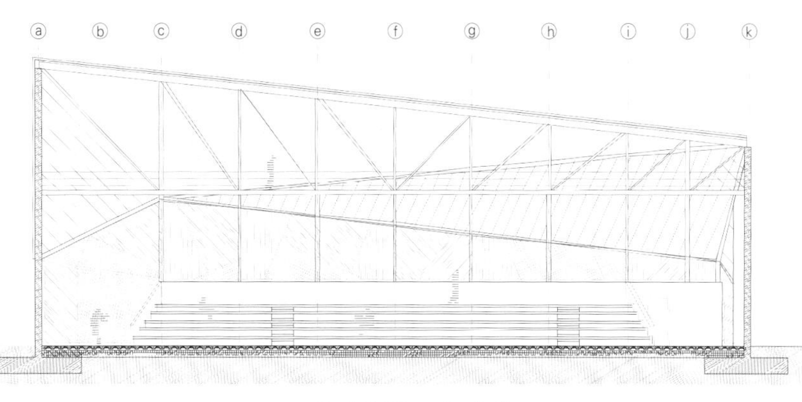

D–D' 剖面图 section D-D'

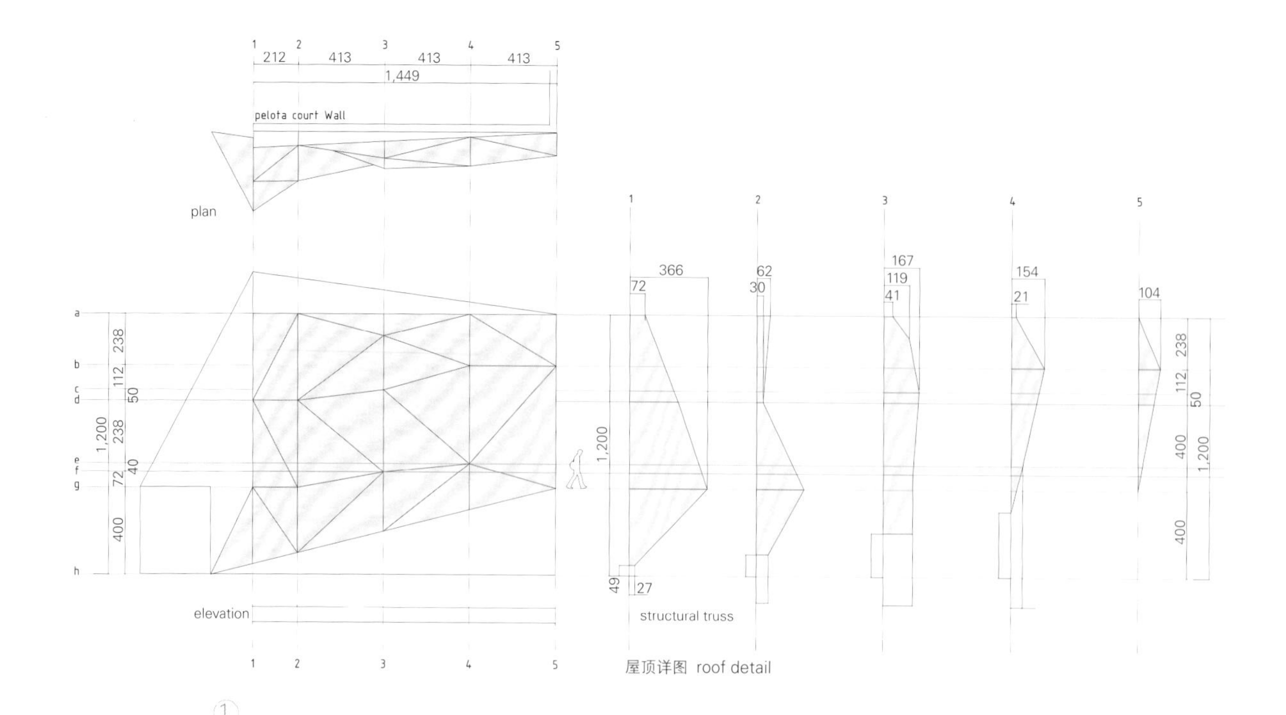

屋顶详图 roof detail

Polycarbonate Fachade,
DANPALON system:

. 16mm thick and 6 layers multicell polycarbonate panel. Colorless, heat sealed in factory.
. Limit formed by aluminum profiles 2AL4 and 2AL5 of the Danpalon system.
. Tied profile: Lacquered steel profile, every 150 cm of length.

HEB 220

Deck section: (outside to inside)
. Quartz-zinc 0,8 mm thick sheet, coil of 650 mm, Standing seam.
. Delta VMZ high density polyethylene film .
. 19mm waterproof, agglomerate board, with tongue and grove joint system.
. Omega profile, made of 40mm galvanized steel, every 60cm of length.
. Zed made of galvanized steel. 1,5mm thick and placed every 60 cm of length.
. Insulation based on semi-rigid 50 mm thick rockwool panel. 70 kg/m3 density.
. Geotextil sheet
. 1mm thick pre-lacquered steel sheet. P106 perforated

Pedestrian path in the park, porous concrete finished, with limestone curb, 2% cross slope

Flooring:

· Selected gravel, 15 cm.
· Coarse concrete, 5 cm.
· Air chamber, based on plastic vault SOLIGLU type, 15cm
· Reinforced concrete floor 15cm thick, quartz polishing finished, mechanical flotation.
· Masterseal 185 primer + Mastertop TC445 Polyurethane painting.

Wall drain:
· Waterproofing paint based on asphalt and rubber.
· Selected gravel.
· Delta Drain Sheet and drain pipe.
· Covered with PRODRAIN type geotextile sheet
Pvc drain pipe ϕ 125 mm.

详图1 detail 1

项目名称：Park & Basque Pelota Court
地点：Gipuzkoa, Spain
建筑师：VAUMM
技术建筑师：Julen Rozas Elizalde
结构工程师：Landabe
客户：Ayuntamiento de Pasaia
面积：8,300m² / 建筑面积：8,300m² / 预算：1 Mill €
竞赛时间：2012.7 / 竣工时间：2015.7
摄影师：©Aitor Ortiz (courtesy of the architect)

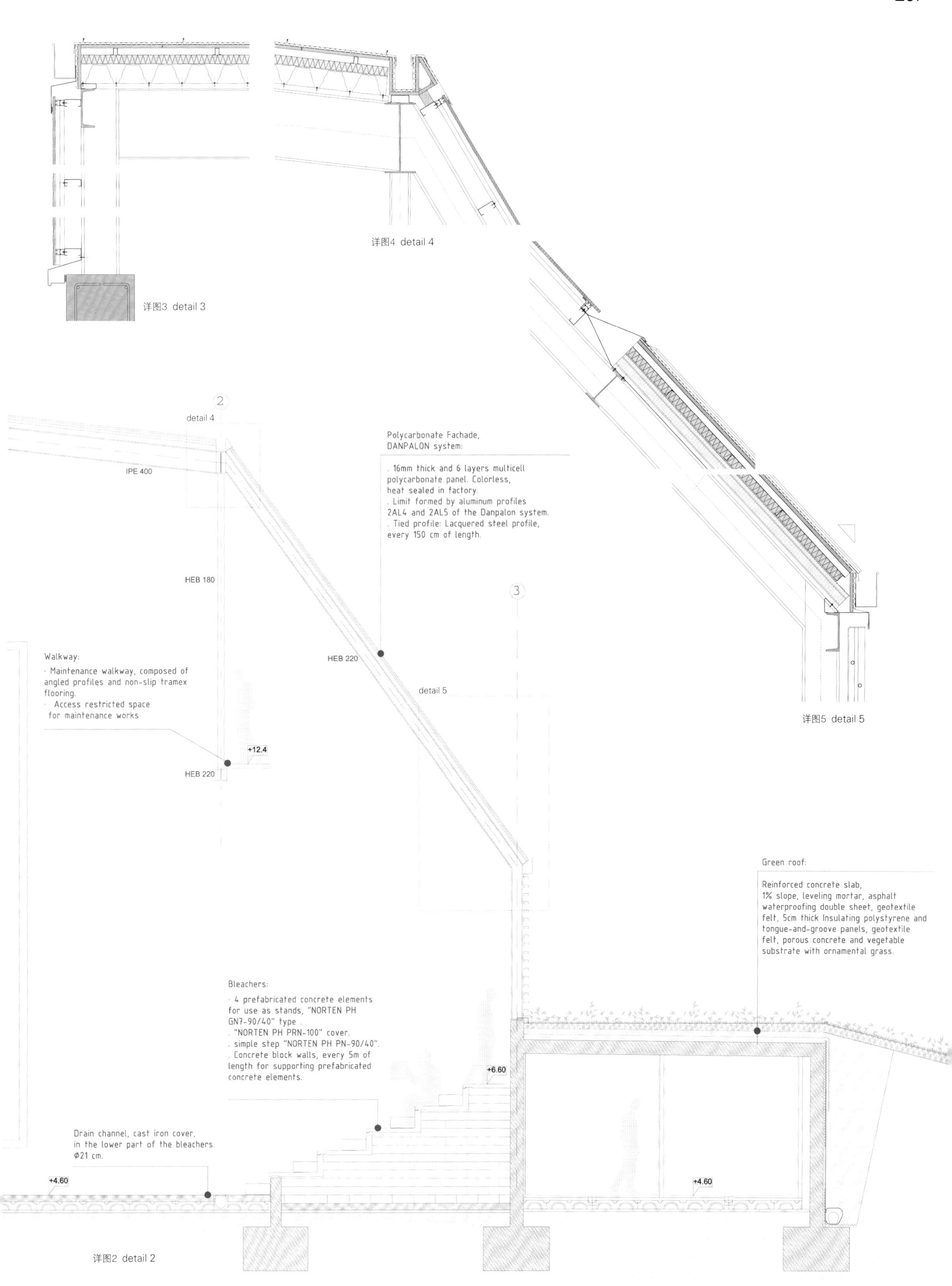

详图3 detail 3
详图4 detail 4
详图5 detail 5
2
detail 4
IPE 400
Polycarbonate Fachade,
DANPALON system:
. 16mm thick and 6 layers multicell polycarbonate panel. Colorless, heat sealed in factory.
. Limit formed by aluminum profiles 2AL4 and 2AL5 of the Danpalon system.
. Tied profile: Lacquered steel profile, every 150 cm of length.
HEB 180
3
Walkway:
· Maintenance walkway, composed of angled profiles and non-slip tramex flooring.
· Access restricted space for maintenance works
HEB 220
detail 5
+12.4
HEB 220
Green roof:
Reinforced concrete slab, 1% slope, leveling mortar, asphalt waterproofing double sheet, geotextile felt, 5cm thick Insulating polystyrene and tongue-and-groove panels, geotextile felt, porous concrete and vegetable substrate with ornamental grass.
Bleachers:
· 4 prefabricated concrete elements for use as stands, "NORTEN PH GN7-90/40" type .
. "NORTEN PH PRN-100" cover.
. simple step "NORTEN PH PN-90/40".
. Concrete block walls, every 5m of length for supporting prefabricated concrete elements.
+6.60
Drain channel, cast iron cover, in the lower part of the bleachers. Φ21 cm.
+4.60
+4.60

详图2 detail 2

萨拉曼卡市体育馆

Carreño Sartori Arquitectos

由于经济状况的原因，这座位于萨拉曼卡市，设计于 2007 年的体育设施在九年之后的 2016 年才得以竣工。然而，正是在这意想不到的延期期间，该项目能够基于对现场更为准确的实际观测，对设计进行具体的修改。因此，那些在项目中发挥关键作用的最初想法能够有机会最终成为现实。

建筑师在设计过程中考虑到了场地特殊的城市环境，同时也考虑到体育馆要能够容纳 2000 名观众。场地的旁边是一个小山，包括一个旧的体育馆、一个足球场和一个废弃的游泳池。虽然归市政府所有，但这些空间都是密闭的，使用起来更像是私人财产。它们不仅彼此独立，也较为分散。因此这个项目的主要目标之一就是将它们连为一体。为此，建筑师为公众设计了一个带有新型体育馆的运动广场，将几何形状的建筑划分出不同的空间界限。因此，不同的路径和入口将具有不同功能的空间连在一起。

考虑到建筑周围的环境，建筑师在每个立面上都做了很独特的设计，与环境相呼应。建筑的南立面面向城市，这儿就成为所有设施的主入口。而由于北立面受到阳光直射，北面的空间通常在白天使用，体育

馆、咖啡厅和行政办公室等就位于北面。在建筑的东面，一条入口坡道与街上的人行步道相连，这样就方便东面的人们直达北门入口。而与此对应的另一面，西面，有通向足球场的应急通道和辅助空间。这里有一个记者房间和公共更衣室，为户外足球场提供支持服务。同时，该设计在汇纳原有的路径和空间的同时，也完全向周围人行步道敞开。

新体育馆通过其复杂的几何结构空间设计为公众开辟出一个独特的外部空间，想要实现这种设计，还是存在一定的困难的。然而，建筑师们还是想出一个办法，将支撑最小化，消除垂直障碍，进而得到斜切面和因此形成的开放空间，最终创造出一个可以供很多人在此聚集的正面空间。这个空间毗邻南入口，上面是一个巨大的倾斜式屋顶。斜面可以用来遮阳和挡雨，人们也可以在此感受附近小山吹来的徐徐清风。新建的广场与一条人行步道连为一体，而这条人行步道又与城市的人行道路系统相连。为了让公众能重新利用废弃的游泳池，建筑师得到市政府批准，使这一切设计成为可能。

由于体育馆也经常举行各种会议和公共活动，因此建筑师也要考虑室内空间间接可调控灯光的设计。为此，建筑师设计了木屋顶，木板折叠排放，形成间隙开口。具有折叠效果的平面的位置依据太阳运行的轨迹和建筑物辅助空间的作用而设计。另外，该建筑场地的地平面有些倾斜，这个抬升建筑体量的多面几何体外观与其所处位置的地势相呼应，和谐共鸣。

从木材到钢材，从钢材到混凝土，该项目遵循一种结构逻辑，运用多种不同元素进行力的传递。这种力的层层传递设计不仅能欺骗人的眼睛，让人们误以为各个平面和空间是漂浮在空中的，同时还可以腾出地面空间用作公共空间。

Municipal Gym of Salamanca

Due to economic circumstances, this sport facility located in Salamanca was completed in 2016, nine years after its design in 2007. However, during the unexpected period of postponement, the project was able to apply specific modifications based on close observations of the site. Therefore, the initial ideas, playing important roles in the project, had the chance to develop into the final result.

The specific urban conditions of the site were considered for the design of the building and its 2,000 spectators. Next to a hill, the site included an old gym along with a soccer field and an abandoned pool. Although owned by the municipality, these spaces were enclosed and used like private properties. They were not only disconnected with each other, but also detached from the site's boundaries, scattered and isolated. Therefore, associating them into an integrated facility was one of the main goals of the project. For this, a sports esplanade is proposed for the public with a new gym that generates spatial boundaries in-between through its geometrical form. Thus the spaces with different purposes are now able to be connected by different paths and accesses.

In relation with the surrounding condition, each facade of the project proposes a distinctive architectural response. The south facade which faces the city, becomes the main access for most of the facilities. On the other hand, the north facade receiving direct sunlight incorporates spaces for daily usages,

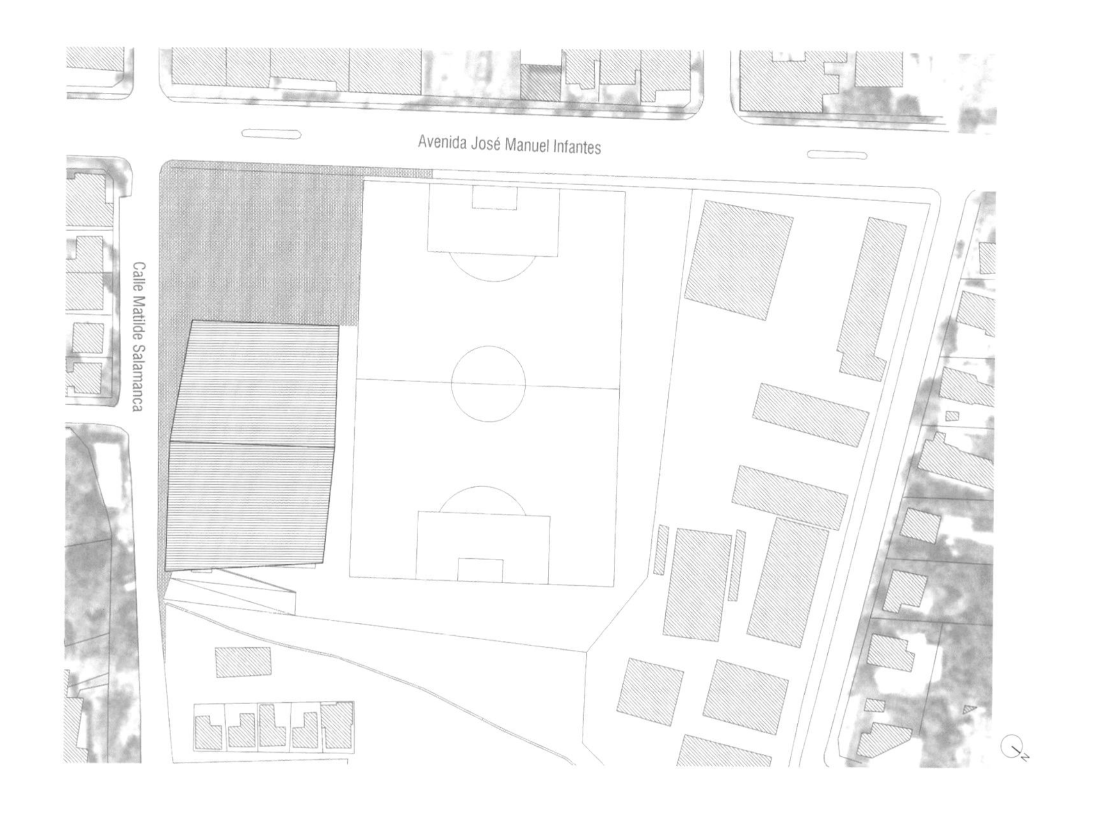

Avenida José Manuel Infantes
Calle Matilde Salamanca

西南立面 south-west elevation

东南立面 south-east elevation

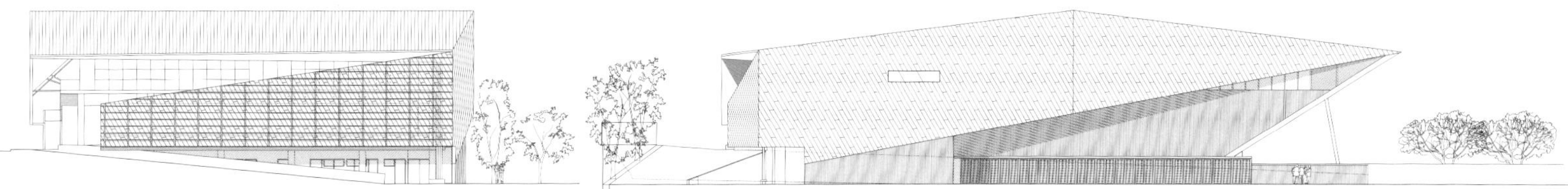

东北立面 north-east elevation

西北立面 north-west elevation

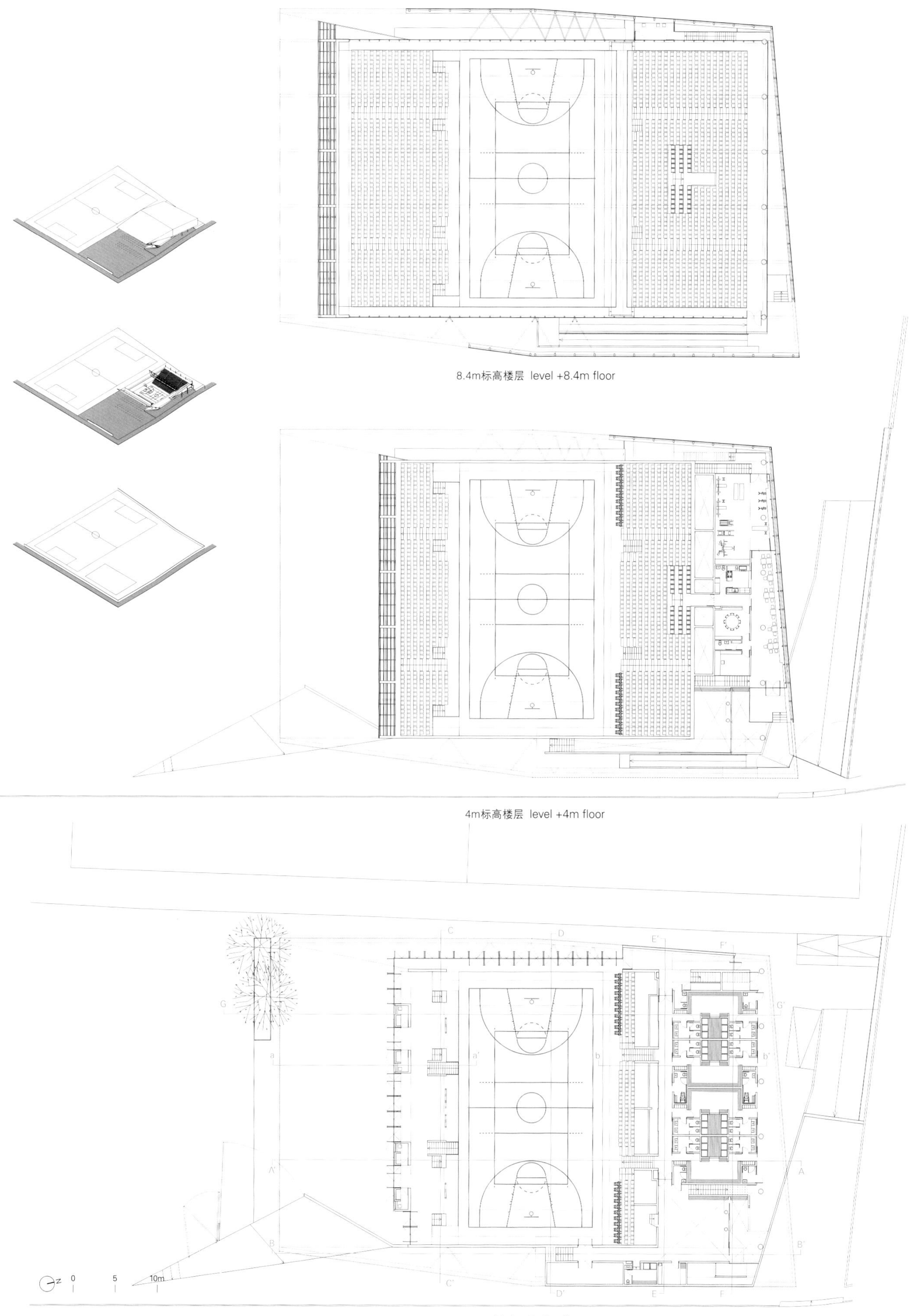

8.4m标高楼层 level +8.4m floor

4m标高楼层 level +4m floor

一层 level +0m floor

such as the gym, cafe, and administrative offices. Towards the east, an access ramp joins the pedestrian walkway of the street so that a direct path is laid to the northern entrances. Opposite to this side of the building, there are complementary spaces and emergency exists leading to the soccer field. Here, a reporter's room and shared dressing rooms provide supportive functions for the outdoor arena. Meanwhile, the site completely opens to the surrounding pedestrian walkways while incorporating its original paths and spaces.
The new gym proposes a unique exterior space for the public through its complex geometry and structure. Because of this, there were certain difficulties in realizing this design feature. However, the project found a way to minimize support and eliminate vertical obstruction to maintain the diagonal plane and its subsequent open space. In result, a frontal space where large number of people can gather freely was created. Adjacent to the main south access, this empty space is covered by an enormous slanting roof. Here the diagonal elevation provides shadow and rain protection, while also opening up to breezes from the nearby hills. Also, the new square is incorporated into an esplanade that connects with the urban pedestrian system. This was possible through the municipality's approval of repurposing the abandoned pool for the public.
As meetings and public events are also held in the gym, indirect and controlled light was considered for the interior space. This was achieved with a wooden roof that folds and creates interstitial openings. The folded planes are set in relation of the sun's direction and complementary spaces of the building. Also, the faceted geometry of the lifted volume resonances the inclined horizontal ground planes in which the building is inserted.
From wood to steel, and from steel to concrete, the project follows a structural logic that delivers loads through different elements. This displacement of forces not only deceives the eye into seeing floating planes and spaces, but also clears out the ground floor for the public.

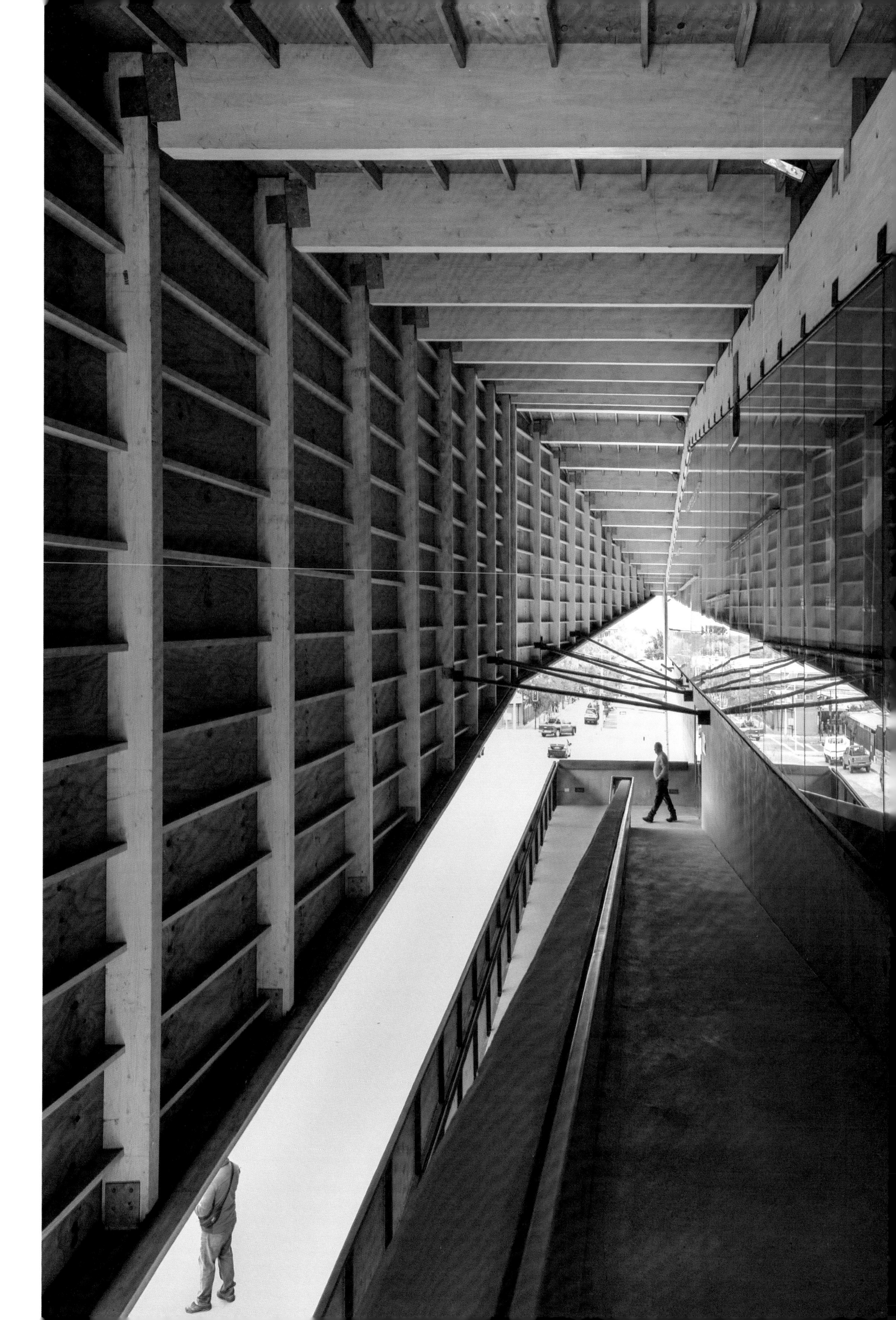

periodo
fouls t.o.l jugador t.o.l fouls
tantos juego tantos

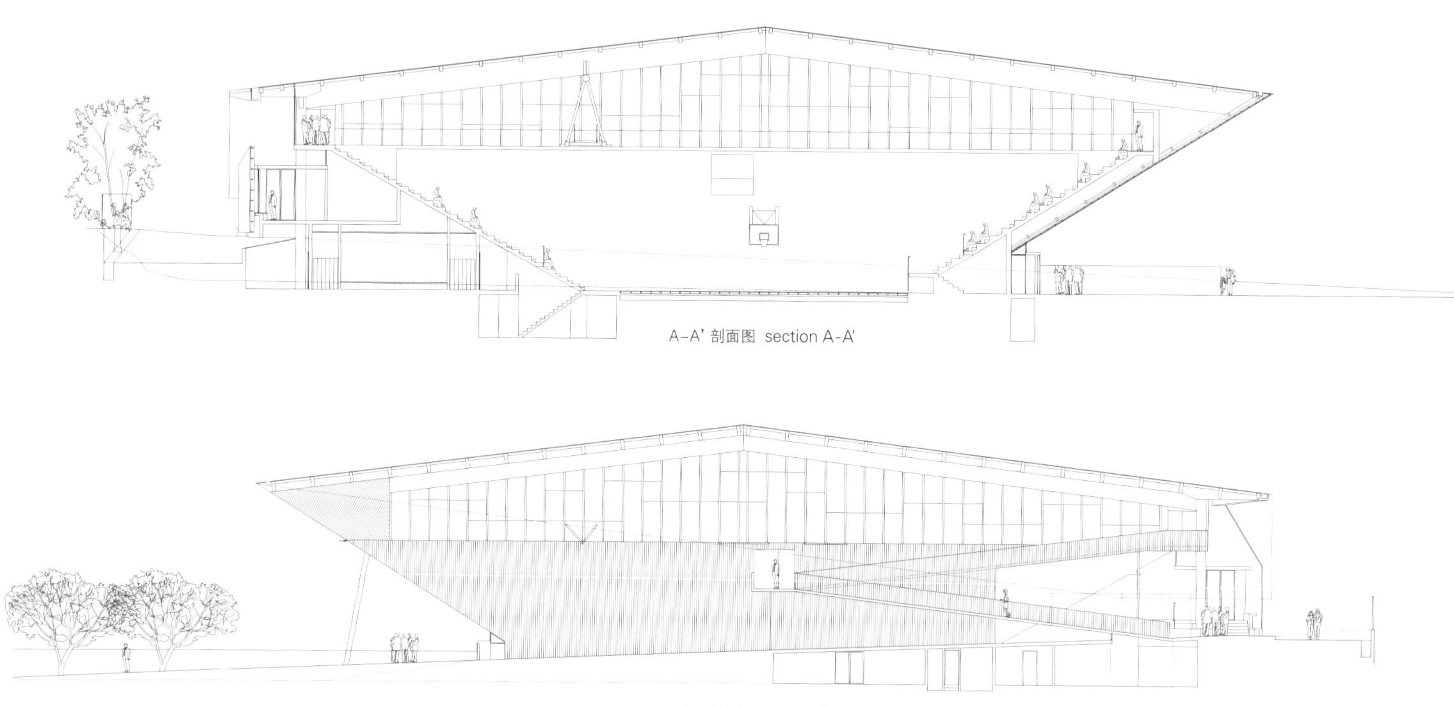

A-A' 剖面图 section A-A'

B-B' 剖面图 section B-B'

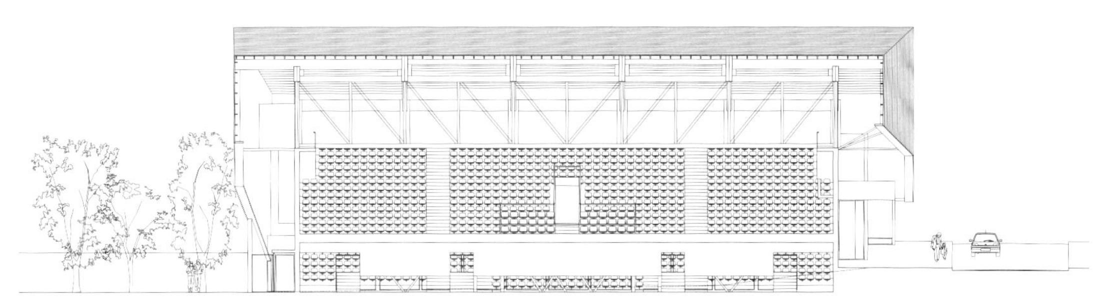

C-C' 剖面图 section C-C'

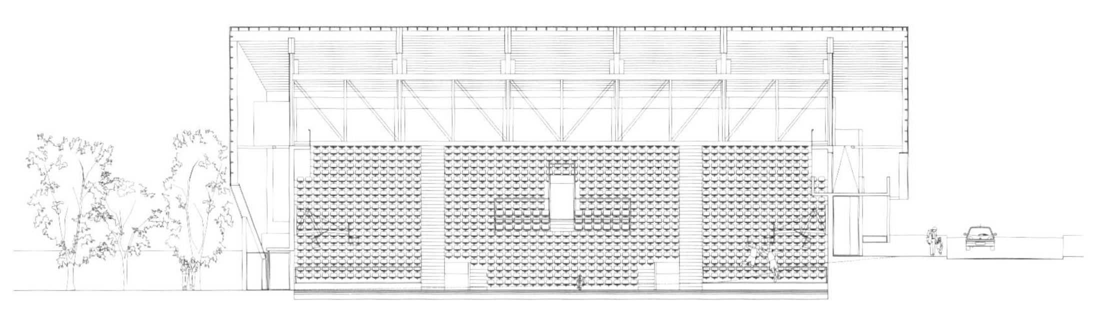

D-D' 剖面图 section D-D'

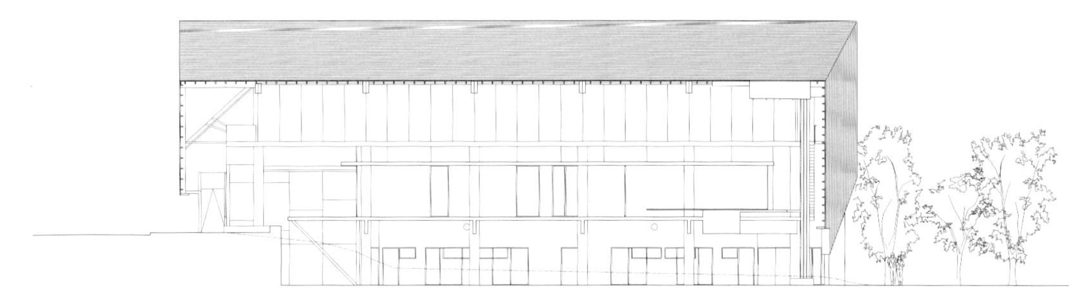

E-E' 剖面图 section E-E'

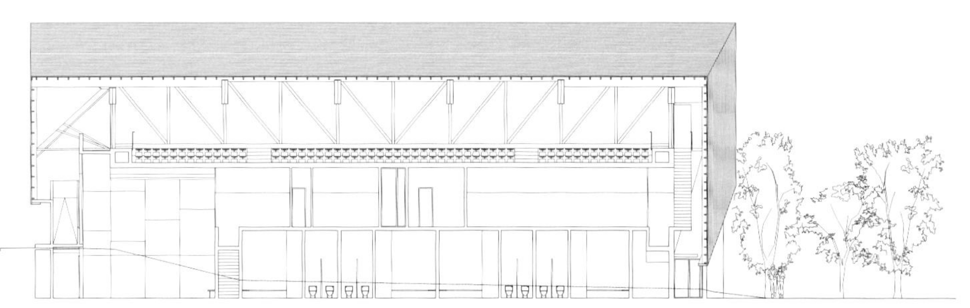

F-F' 剖面图 section F-F'

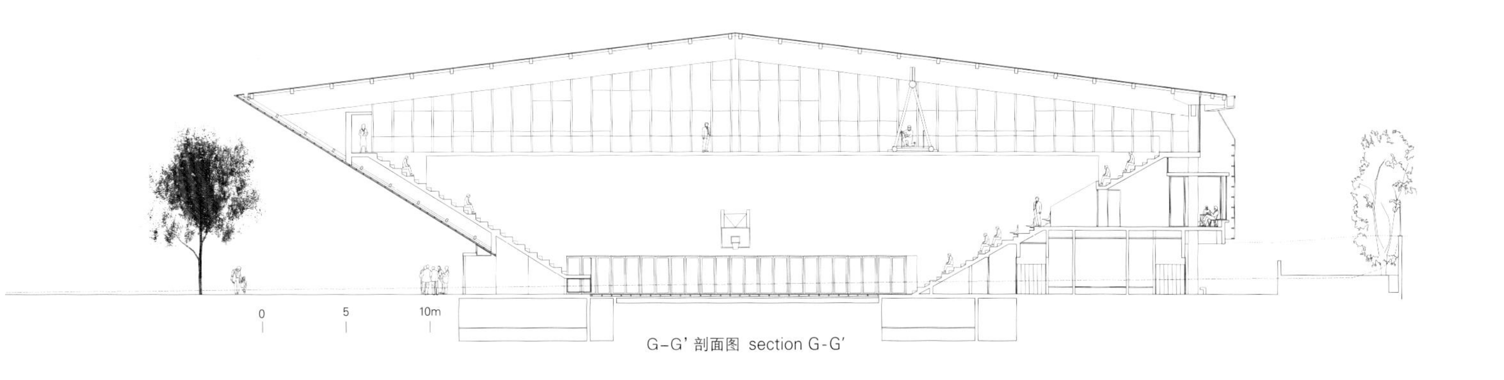

G-G' 剖面图 section G-G'

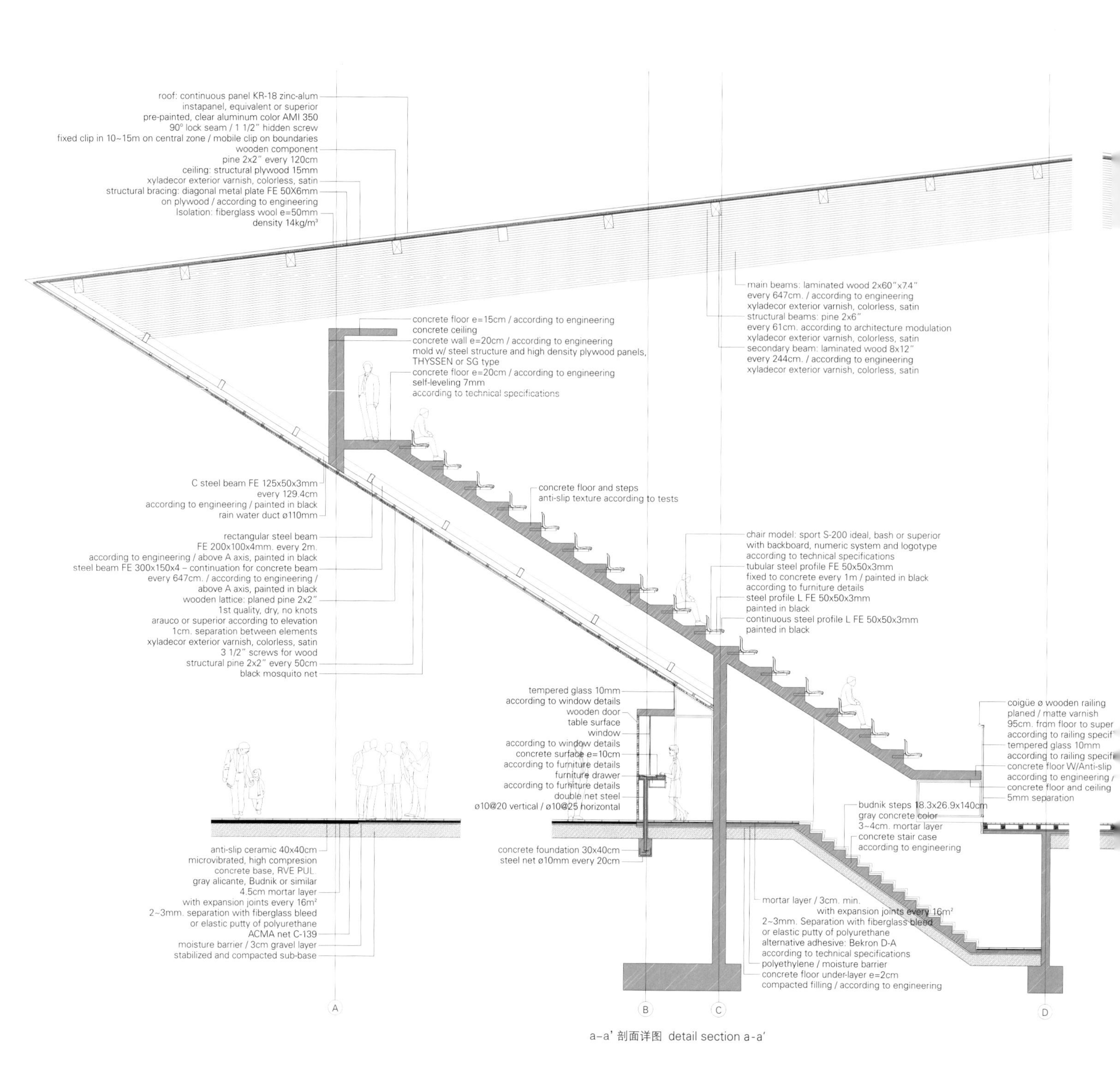

a–a' 剖面详图 detail section a-a'

项目名称：Gimnasio Municipal De Salamanca
地点：Salamanca, Chile
建筑师：Mario Carreño Zunino, Piera Sartori del Campo_Carreño Sartori Arquitectos
合作者：Pamela Jarpa Rosa, Pia Mastrantonio Pedrina, Claudia Wagner Böhmer
结构计算：SyS Ingenieros Consultores
电气：ICG S.A.
卫浴设施：Roberto Pavéz Mujica
地形设计：Ángel Espinoza Varas
景观设计：Carreño Sartori Arquitectos
施工：Constructora Lohse y Villablanca ltda.
用地面积：14,010m² / 建筑面积：3,257.9m²
设计时间：2007 / 施工时间：2009—2016 / 竣工时间：2016
摄影师：©Marcos Mendizabal (courtesy of the architect)

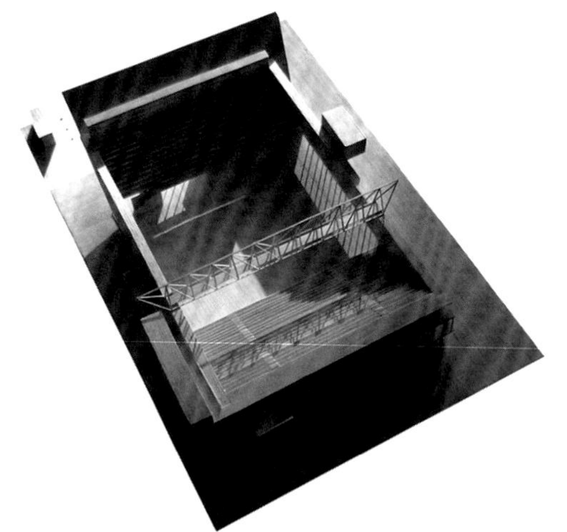

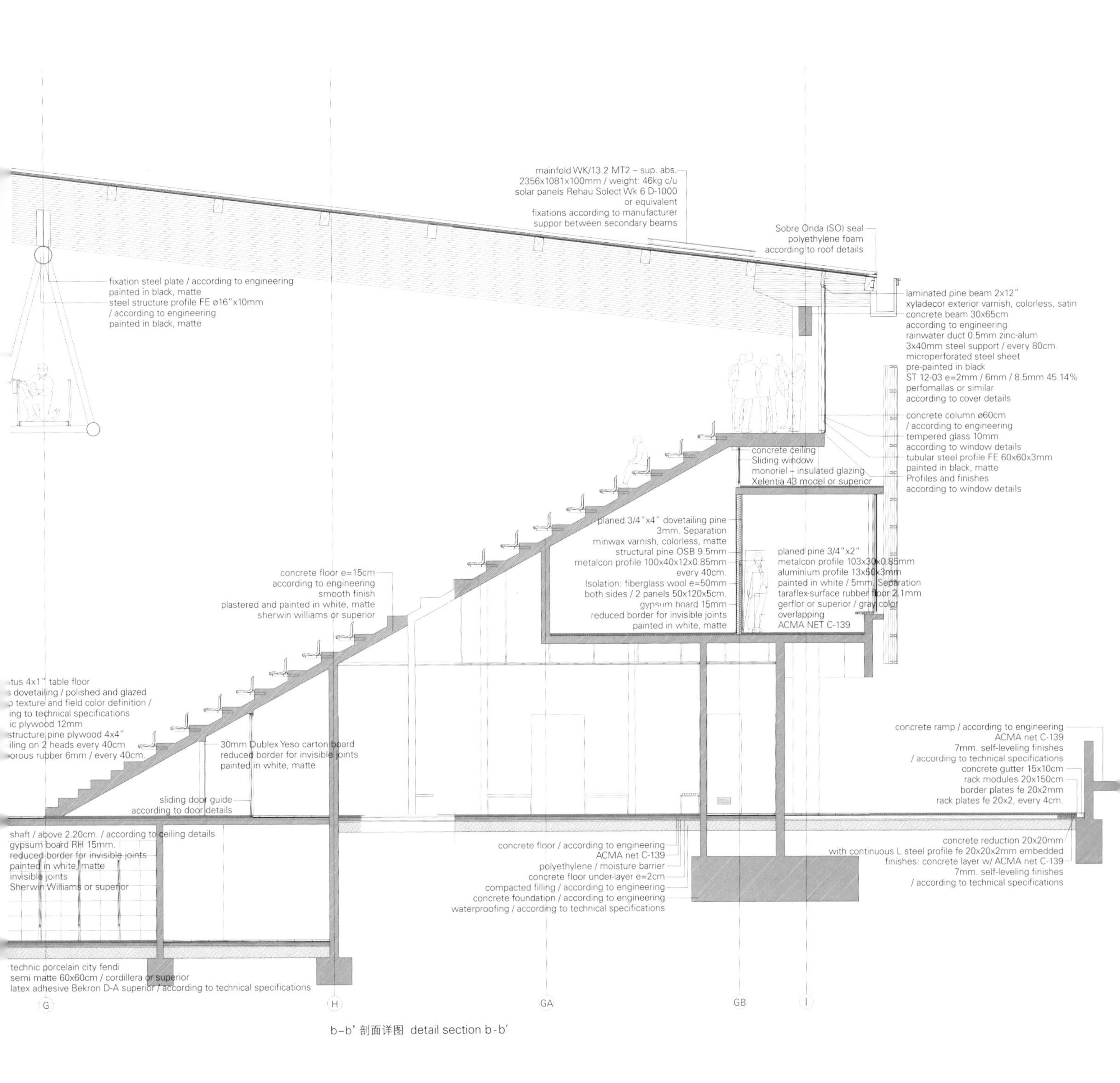

b–b' 剖面详图 detail section b-b'

P104 Studio Fuksas

Led by Massimiliano Fuksas[right] and Doriana Mandrelli Fuksas[left], laureates of Commandeur de l'Ordre des Arts et des Lettres de la République Française. They have worked in Europe, Africa, America, Asia and Australia for over the past 40 years with headquarters in Rome, Paris and Shenzhen, and a staff of 170 professionals. Massimiliano Fuksas was born in Rome, 1944 and graduated from the University of Rome "La Sapienza" in 1969. In 1999 he received the Grand Prix National d'Architecture Française. In 2010 he was decorated with Légion d'Honneur by the French President. Was given an Honorary Fellowship from the AIA, RIBA . Is member of the Académie d'Architecture in Paris. In 2006 he was named Cavaliere di Gran Croce della Repubblica Italiana. Has been Visiting Professor the Columbia University in New York, the École Spéciale d'Architecture in Paris, the Akademie der Bildenden Künste in Wien, the Staatliche Akademie der Bildenden Künste in Stuttgart. Doriana Fuksas was born in Rome and graduated in History of Modern and Contemporary Architecture at the University of Rome "La Sapienza" in 1979. Earned a degree in Architecture from ESA, École Spéciale d'Architecture, Paris. Has worked with Massimiliano Fuksas in 1985 and has been director in charge of "Fuksas Design" since 1997.
They were author of the Design column in the Italian newspaper "La Repubblica" in 2014-2015.

P14 DVVD

s a Paris based group of architects, engineers, lesigners, builders and thinkers effective within he fields of architecture, urbanism, research and levelopment. The office gathers over forty people, nvolved in a large number of projects throughout Europe. To deal with today and tomorrow challenges n urban design and architecture, they explore new ields by overlapping conventional approaches, aking advantage of their solid know-how, with new concept to better fit the future life forms. Create unique, lively, sustainable projects that reach beyond themselves and become a durable value to he users, the society and the culture they are built nto. Founder and CEO, Daniel Vaniche is a graduate of the Ecole Polytechnique and Ecole Nationale les Ponts et Chaussées engineering schools, and architect. His experience both as an architect and an engineer convinced him that to bring these two skills together within a single structure made sense. Co-founder, Vincent Dominguez has contributed o the experience in engineering, architecture and lesign right from the earliest days, not only regarding design itself but also in terms of expertise, nethods and work-site techniques. His focus is on arge scale public buildings in various cultural contexts. His overview and attention to detail strongly nfluence the office approach.

P150 Studio ARCHATTACKA

Was founded by Andrey Voronov and Alexander Berzing in 2013. Organized based on a young architectural workshop of LLC Architectural Workshop LES (from 2011 to 2013 - TPO Lesosplaw), for participation in various art and architectural tenders, and also for promotion of various projects in the field of the modern art. Conducts a vigorous business activity in the sphere of architectural designing, engaged with both private and public projects.
from left. Andrey Voronov, Alexander Berzing, and Innokenty Padalko

P30 Rey-Lucquet et associés

Can look back on many years of experience in renovation and expansion of buildings. The sports center n Marckolsheim, the College Foch in Haguenau, the hotel complex in Bischoffsheim or the Lycee Bartholdi in Colmar show the versatility and professionalism of the atelier concerning these sensitive and challenging tasks. Lead by Serge Lucquet, Thierry Rey, Olivier de Crecy.

P30 Dietrich | Untertrifaller Architects

Founded in 1994 by Helmut Dietrich and Much Untertrifaller. In 1992, they won the international competition for the conversion and extension of the Festspielhaus (Festival Theatre) in Bregenz, which was delivered in 2006. They are well known for their expertise in timber design and construction. Today, the agency employs an international team of 70 architects working in Austria (Bregenz and Vienna), Switzerland (St Gallen), France (Paris) and Germany (Munich). Is lead by Helmut Dietrich, Much Untertrifaller, Dominik Philipp and Patrick Stremler. Received the Prix d'Architecture de Bretagne 2016, the LEED Platinum and the International Architecture Award.
from left. Maria Megina, Dominik Philipp, Helmut Dietrich, Heiner Walker, Patrick Stremler, Much Untertrifaller, Peter Nussbaumer, Ulrike Bale-Gabriel, Susanne Gaudl, Michael Porath and Christof Staheli

P76 UNStudio

Ben van Berkel studied architecture at the Rietveld Academy in Amsterdam and at the Architectural Association in London, receiving the AA Diploma with Honours in 1987. Set up Van Berkel & Bos Architectuurbureau with Caroline Bos in Amsterdam in 1988. They established UNStudio (United Net) in 1998. UNStudio presents itself as a network of specialists in architecture, urban development and infrastructure. Currently he holds the Kenzo Tange Visiting Professor's Chair at Harvard University Graduate School of Design. With UNStudio he realized amongst others the Mercedes-Benz Museum in Stuttgart, Arnhem Central Station in the Netherlands, the façade and interior renovation for the Galleria Department store in Seoul, the Singapore University of Technology and Design and a private villa up-state New York.

P44 BudCud

Is a contemporary practice from Cracow, Poland), operating within the fields of architecture and urbanism since 2010. Led by Mateusz Adamczyk[right] and Agata Wozniczka[left], often collaborates with professionals of different disciplines. Has worked on offices, commercial interiors, public spaces, exhibitions and urban startegies. Aims in searching for and designing complex architectural environments. Projects are logical and professional elaborations of better and sustainable reality in modern times, designed for the future inhabitants who benefit from a new spatial reality. Its design process is a constant evaluation of a rational model scheme, defined by both context and experimentation. They are heavily influenced by the project environment, which makes BudCud proposals contextually conscious, but not naive. Implementing such a strategy results in something that is both unusual and unique.

P54 Coop Himmelb(l)au

Was founded in Vienna, 1968 and opened another branch in LA, 1988. Wolf D. Prix, co-founder, has been operating the firm as CEO and Design Principal. He was born in 1942, Vienna and studied architecture at the Vienna University of Technology, AA School of London and SCI-Arc in LA. Is a member of the Association of German Architects (BDA), Royal Institute of British Architects (RIBA), and Fellow of the American Institute of Architecture (FAIA). Is also a permanent member of the Austrian Art Senate and the European Academy of Sciences and Arts. In 2015 he was awarded an Honorary Diploma of the Architectural Association of London.

P208 Carreño Sartori Arquitectos

Mario Carreño Zunino, architect graduated from the Pontificia Universidad Católica de Chile (PUCCH) in 2000. Teacher at the Architecture School of the PUCCH since 2003 up to now. Piera Sartori del Campo, architect graduated from the Pontificia Universidad Católica de Chile (PUCCH) in 1999, landscape architect graduated from the PUCCH in 2003. Teacher at the Architecture School of the PUCCH since 2003 up to now. In 2000, they founded Carreño Sartori Architects in Santiago de Chile, developing both public and private architectural and landscape design projects, willing to link architecture to territorial area, connecting materials nature to different constructive systems. Their work has been part of various publications, exhibitions and biennials.

P130 **5+1AA Alfonso Femia Gianluca Peluffo**

Alfonso Femia and Gianluca Peluffo founded 5+1 in 1995 and created 5+1AA in 2005. In 2006, Simonetta Cenci became a partner of 5+1AA and they opened a studio in Milan. They deal with the theme of simultaneity in the relationship between city, territory and architecture, constructing it in the form of reality. In 2011, they won the Philippe Rotthier European Prize for Architecture and the International Chicago Athenaeum Prize. In 2012, they won the Architecture Rivelate Award and the Trimo Architectural Award.

P194 **VAUMM**

Is established in 2002 by Tomás Valenciano Tamayo, Jon Muniategiandikoetxea Markiegi, Javier Ubillos Pernaut, Marta Álvarez Pastor, and Iñigo García Odiaga[from left]. Adding specialists in different disciplines, both within and outside the office, adding basic knowledge to provide solutions. Through this variable organization, the permanent office is enlarged thanks to multiple collaborations depending on the situation, focus on forming each project the best team possible. Recently, they received 1st prize at restricted competition for Dantzagunea-Arteleku-Scenic Arts and Dance & Cultural in Errenteria in 2013 and for the Remodelling of Molinao Park and Sports Center in Pasaia in 2012. Participated in the exibition entitled '20 Spanish offices in Mexico, 20 Mexican offices in Spain'.

Richard Ingersoll

Born in California, 1949, earned a doctorate in architectural history at UC Berkeley, and was a tenured associate professor at Rice University (Houston) from 1986-97. He has lived off and on in Tuscany since 1970 and currently teaches at Syracuse University in Florence (Italy), and the Politecnico in Milan. He was the executive editor of **Design Book Review** from 1983-1997. His recent publications include: **World Architecture. A Cross-Cultural History** (2013); **Sprawltown, Looking for the City on its Edge** (2006); **World Architecture, 1900-2000. A Critical Mosaic, Volume I: North America, USA and Canada** (2000). He frequently writes criticism for **Arquitectura Viva, Architect, Lotus**, and **Bauwelt**.

Isabel Potworowski

Graduated from TU Delft with a Master in Architecture, and currently works for Barcode Architects in Rotterdam. During the graduate studies, Potworowski was a member of the editorial committee and wrote several articles for the independent student journal **Pantheon**. Originally from Canada, She completed her Bachelor in Architecture at McGill University in Montreal, where she was awarded the Louis Robertson book prize for the highest grade in first year history. She has also studied for one semester at the Politecnico di Milano. She has worked at ONPA Architects and Manasc Isaac Architects, both in Edmonton, Canada.

P162 **AZPML**

Is an international architecture practice with offices in London, Zurich, and Princeton. Was founded in 2011 by Alejandro Zaera-Polo[right] and Maider Llaguno[left]. Alejandro Zaera-Polo was born in 1963 and studied at the ETSA in Madrid, graduating with Honors, and received a masters degree from Harvard GSD in 1991 with Distinction. Was Dean of the School of Architecture at Princeton University, Berlage Institute in Rotterdam. Was also Berlage Chair in the TU Delft. Maider Llaguno studied at the ETSA in San Sebastian and Barcelona. Received her Diploma in Architecture with Honours in 2006. Graduated with honours in Advanced Architecture Design from The GSAPP, Columbia University in 2010.

图书在版编目(CIP)数据

公共建筑改造 / 意大利福克萨斯建筑设计事务所等编 ; 王京等译. — 大连 : 大连理工大学出版社, 2017.11
(建筑立场系列丛书)
ISBN 978-7-5685-1112-4

Ⅰ. ①公… Ⅱ. ①意… ②王… Ⅲ. ①公共建筑—技术改造 Ⅳ. ①TU242

中国版本图书馆CIP数据核字(2017)第265905号

出版发行:大连理工大学出版社
(地址:大连市软件园路80号　邮编:116023)
印　　刷:上海锦良印刷厂
幅面尺寸:225mm×300mm
印　　张:14.25
出版时间:2017年11月第1版
印刷时间:2017年11月第1次印刷
出 版 人:金英伟
统　　筹:房　磊
责任编辑:杨　丹
封面设计:王志峰
责任校对:张昕焱
书　　号:978-7-5685-1112-4
定　　价:258.00元

发　行:0411-84708842
传　真:0411-84701466
E-mail:12282980@qq.com
URL: http://dutp.dlut.edu.cn

本书如有印装质量问题，请与我社发行部联系更换。